The Facts On File

DICTIONARY
of
PHYSICS

The Facts On File

DICTIONARY
of
PHYSICS

Edited by
John Daintith
Consultant Editor
Eric Deeson

Facts On File, Inc.
119 West 57th Street
New York, N.Y. 10019

The Facts On File
Dictionary of Physics
Edited by John Daintith
Consultant Editor Eric Deeson

First edition published in the United Kingdom in 1981 by
Intercontinental Book Productions Limited, Berkshire House,
Queen Street, Maidenhead, Berkshire, SL6 1NF England, in
conjunction with Seymour Press Ltd.

Second edition published in the United States of America in 1981
by Facts On File, Inc., 119 West 57th Street, New York, N.Y. 10019

Library of Congress Cataloging in Publication Data

Daintith, John; Deeson, Eric.
　The Facts on File dictionary of physics.

　1. Physics—Dictionaries.　I. Facts on File,
inc., New York.　II. Title.　III. Title: Dictionary
of physics.
QC5.D33　　530′.03′21　　80-26854
ISBN 0-87196-511-9

Printed in Singapore and the U.S.A. by Murray Printing Co.,
Forge Village, Mass.

The illustration on the jacket of this book is a photograph of
radiation diffusion.

ACKNOWLEDGEMENTS

Contributions
Roger Adams B.Sc.
Jane Craig B.Sc. A.R.C.S.
Eric Deeson M.Sc. F.C.P. F.R.A.S.
Sue Flint B.Sc.
B. W. Lowthian B.Sc. Ph.D.
Michael S. Slater B.Sc.
James M. Struthers B.Sc.
M. Welton B.Sc.
Carol Russell B.Sc.

Computer systems
Barry Evans

Production editors
Sarah Mitchell
Elizabeth Tootill

A

ab- A prefix used with a practical electrical unit to name the corresponding electromagnetic unit. For example, the electromagnetic unit of charge is called the *abcoulomb. Compare* stat-.

aberration A defect in an optical system such that the image is not a true picture of the object. For instance, coloured fringes may appear, the image may not be equally focused, or the shape may show distortion. Techniques of aberration correction exist; these can, however, be complex and costly.

Chromatic (colour) *aberration* is found with a single lens; mirrors do not suffer from chromatic aberration. Because dispersion always accompanies refractive deviation, the 'red' image will be further from the lens than the 'blue'. Consequently, the image is surrounded by coloured fringes. Chromatic aberration is corrected by forming a compound lens, whose elements have different refractive constants.

Spherical aberration always occurs with rays that are distant from the axis and incident on a spherical mirror or lens. It is the cause of the caustic curve. Spherical aberration is corrected by using parabolic reflecting and refracting surfaces.

Astigmatism affects rays neither close nor parallel to the axis. The cone of rays through a lens from an off-axis object does not focus at a point. Instead, two images in the form of short lines are formed at different distances from the lens. Between the two the image appears circular. Mirrors forming images of off-axis points show a similar defect. The best method of minimizing astigmatism is to reduce the aperture with stops, thus allowing light only through the centre of the lens.

Coma is rather similar in cause, effect, and correction to astigmatism. After refraction by a lens, a cone of rays from an off-axis object tends to have a tadpole-shaped section because of coma. *Distortion* is the result of differences in a lens' magnifying power between different axes. Reduction of aperture is the normal solution to both coma and distortion.

absolute expansion *See* expansivity.

absolute humidity The mass of water vapour per unit volume of air, usually measured in kilograms per cubic metre. *Compare* relative humidity. *See also* humidity.

absolute permeability *See* permeability.

absolute refractive constant *See* refractive constant.

absolute temperature Symbol: T A temperature defined by the relationship $T = \theta + 273.15$, where θ is the Celsius temperature. The absolute scale of temperature was a fundamental scale based on Charles' law applied to an ideal gas:
$$V = V_0(1 + \alpha\theta)$$
where V is the volume at temperature θ, V_0 the volume at 0, and α the thermal expansivity of the gas. At low pressures (when real gases show ideal behaviour) α has the value $1/273.15$. Therefore, at 0 = -273.15 the volume of the gas theoretically becomes zero. In practice, of course, substances become solids at these temperatures. However, the extrapolation can be used for a scale of temperature on which -273.15°C corresponds to 0° (absolute zero). The scale is also known as the *ideal-gas scale*; on it temperature intervals were called *degrees absolute* (°A) or *degrees Kelvin* (°K), and were equal to the Celsius degree. It can be shown that the absolute temperature scale is identical to the thermodynamic temperature scale (on which the unit is the kelvin).

absolute zero The zero value of thermodynamic temperature; 0 kelvin or -273.15°C.

absorptance Symbol: α The ratio of the radiant or luminous flux absorbed by a body or material to the incident flux. It was formerly called the *absorptivity*.

absorption A process in which a gas is taken up by a liquid or solid, or in which a liquid is taken up by a solid. In absorption, the substance absorbed goes into the bulk of the material. Solids that absorb gases or liquids often have a porous structure. The absorption of gases in solids is sometimes called *sorption*. *Compare* adsorption.

absorption coefficient *See* Lambert's laws.

absorption of radiation No medium transmits radiation without some energy loss. This loss of energy is called absorption. The energy is converted to some other form within the medium. *See also* Lambert's laws.

absorption spectrum *See* spectrum.

absorptivity *See* absorptance.

abundance 1. The relative amount of a given element amongst others; for example, the abundance of oxygen in the Earth's crust is approximately 50% by weight.
2. The amount of a nuclide (stable or radioactive) relative to other nuclides of the same element in a given sample. The *natural abundance* is the abundance of a nuclide as it occurs naturally. For instance, chlorine has two stable isotopes of masses 35 and 37. The abundance of ^{35}Cl is 75.5% and that of ^{37}Cl is 24.5%. For some elements the abundance of a particular nuclide depends on the source.

a.c. *See* alternating current.

acceleration Symbol: a The rate of change of speed (a scalar) or of velocity (a vector). The basic SI unit is the metre per second per second (m s^{-2}). For constant acceleration:

$$a = (v_2 - v_1)/t$$

v_1 is the speed or velocity when timing starts; v_2 is the speed or velocity after time t. (This is one of the equations of motion.) Negative values of a relate to cases of retardation, or deceleration (slowing down).

The equation gives the mean acceleration during the time interval. If acceleration is not uniform (constant), $a = dv/dt$. If the vector form is being used, acceleration means either rate of change of speed in a given direction or rate of change of velocity, including change of direction. The acceleration of an object depends on the net outside force F acting. From Newton's second law, $F = ma$, where m is the object's mass.

acceleration due to gravity *See* acceleration of free fall.

acceleration of free fall (acceleration due to gravity) Symbol: g The constant acceleration of a mass falling freely (without friction) in the Earth's gravitational field. g is a measure of gravitational field strength — the force on unit mass. The force on a mass m is its weight W, where $W = mg$.

The value of g varies with distance from the Earth's surface. Near the surface it is just under 10 metres per second per second (9.806 65 m s^{-2} is the standard value). It varies with latitude, partly because the Earth is not perfectly spherical (it is flattened near the poles).

accelerator A device for accelerating charged particles to high energies so that they are able to penetrate to the nuclei of atoms in a target, causing nuclear reactions. The earliest accelerator was invented by Cockcroft and Walton and was first used to accelerate protons towards a target of lithium.

Two types are now in use. In *linear accelerators* the particles are accelerated in a straight line. *Cyclic accelerators* use magnetic fields to keep the particles moving in circular or spiral paths. Examples of cyclic accelerators are the cyclotron, the synchrocyclotron, and the betatron. *See also* linear accelerator.

acceptor *See* semiconductor.

acceptor circuit *See* resonance.

accommodation The action of the eye in changing its focal power. The normal eye has a high power (short focal distance) for viewing close objects; it relaxes to low power for very distant objects. Accommodation is accomplished by muscles in a ring round the lens of the eye, which are able to change the shape of the lens. The amplitude of accommodation decreases with age — the power range is around 11 dioptres at age 10 and 1 dioptre at age 70. Thus the distance between far point and near point decreases with age. This effect is presbyopia.
See also eye, amplitude of accommodation.

accumulator (secondary cell, storage battery) An electric cell or battery that can be charged by passing an electric current through it. The chemical reaction in the cell is reversible. When the cell begins to run down, current in the opposite direction will convert the reaction products back into their original forms. The most common example is the lead–acid accumulator, used in vehicle batteries.

achromat An achromatic lens.

achromatic colour A colour that has no hue; i.e. black, white, or grey.

achromatic lens A compound lens whose elements differ in refractive constant in order to minimize chromatic aberration. Simple *achromatic doublets* are formed by combining two lenses of different glass. The condition for achromatism is:
$$\omega_1 P_1 + \omega_2 P_2 = 0$$
where ω_1 and ω_2 are the dispersive powers of the glasses of the lenses, and P_1 and P_2 are the powers of the lenses. Achromatic lenses are corrected for chromatic aberration at two different wavelengths. *See also* apochromatic lens.

aclinic line (magnetic equator) *See* isoclinic line.

acoustics The study of the production and properties of sounds. The term is also used to describe the way in which sound is reproduced in practical situations.

actinic radiation Radiation that can cause a chemical reaction; for example, ultraviolet radiation is actinic.

actinometer An instrument for measuring the intensity of radiation.

actinon *See* emanation.

action 1. An out-dated term for force. *See* reaction.
2. The product of energy and time. The Planck constant was originally known as Planck's constant of action.

activated charcoal *See* charcoal.

activity Symbol: A For a radioactive substance, the average number of atoms disintegrating per unit time.

acuity, visual The ability of the eye to see separately two points close to each other. It is a measure of the resolving power of the eye's optical system and depends on the density of cells in the retina. The maximum acuity of the normal human eye is around 0.5 minutes of arc — points separated by this angle at the eye should be seen as separate. *See* resolution.

additive process A process of colour mixing by addition. *See* colour.

adiabatic change A change during which no energy enters or leaves the system.
In an adiabatic expansion of a gas, mechanical work is done by the gas as its volume increases and the gas temperature falls. For an ideal gas undergoing a reversible adiabatic change it can be shown that
$$pV^\gamma = K_1$$

3

adiabatic demagnetization

$$T^\gamma p^{1-\gamma} = K_2$$
and $TV^{\gamma-1} = K_3$

where K_1, K_2, and K_3 are constants and γ is the ratio of the principal specific heat capacities. *Compare* isothermal change.

adiabatic demagnetization A method of producing temperatures close to absolute zero. A sample of paramagnetic salt is cooled in liquid helium in a strong magnetizing field. The field is removed, demagnetizing the sample and cooling it further.

admittance Symbol: Y The reciprocal of impedance, measured in siemens (S). It is a measure of the response of an electric circuit to an alternating signal. *See also* impedance.

adsorption A process in which a layer of atoms or molecules of one substance forms on the surface of a solid or liquid. All solid surfaces take up layers of gas from the surrounding atmosphere. The adsorbed layer may be held by chemical bonds (*chemisorption*) or by weaker van der Waals forces (*physisorption*). *Compare* absorption.

advanced gas-cooled reactor *See* gas-cooled reactor.

aerosol A dispersion of small particles of solid or droplets of liquid in a gas.

aether *See* ether.

agate A crystalline form of silica used, because of its hardness, in making knife edges in balances, pendulums, etc.

agonic line *See* isogonic line.

AGR Advanced gas-cooled reactor. *See* gas-cooled reactor.

air wedge An arrangement producing localized interference patterns by reflection at the two sides of a wedge-shaped film of air (as between two glass slides at an angle). Newton's rings (variable wedge angle) and thin films (zero wedge

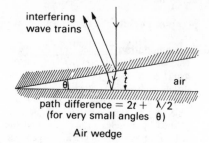

interfering wave trains

path difference $= 2t + \lambda/2$
(for very small angles θ)

Air wedge

angle) produce similar effects. In an air wedge the fringes (with monochromatic light) are light and dark bands parallel to the thin edge of the wedge. A bright fringe occurs when $2t + \lambda/2 = m\lambda$, t being the thickness and m an integer. For a bright fringe $2t = m\lambda$. *See also* interference.

albedo The ratio of the amount of light reflected from a surface to the amount of incident light.

alcohol thermometer A liquid-in-glass thermometer that uses ethanol as its working substance. The ethanol commonly contains a red dye to make the liquid more visible. *See also* thermometer.

allotropy The existence of a solid substance in different physical forms. Tin, for example, has metallic and non-metallic crystalline forms. Carbon has two crystalline allotropes: diamond and graphite.

alloy A mixture of two or more metals (e.g. bronze or brass) or of a metal with small amounts of non-metals (e.g. steel). Alloys may be completely homogeneous mixtures or may contain small particles of one phase in the other phase.

alpha decay A type of radioactive decay in which the unstable nucleus emits a helium nucleus. The resulting nuclide has a mass number decreased by 4 and a proton number decreased by 2. An example is:

$$^{226}_{88}Ra \rightarrow ^{222}_{86}Rn + ^{4}_{2}He$$

The particles emitted in alpha decay are *alpha particles*. Streams of alpha particles are *alpha rays* or *alpha radiation*. They are less penetrating than beta particles. *See also* beta decay.

alternating current (a.c.) Electric current that regularly reverses its direction. In the simplest case, the current varies with time (t) in a simple harmonic manner, represented by the equation $I = I_0\sin2\pi ft$, f being the frequency. Alternating current can be described by its peak value I_0, or by its root-mean-square value I_{RMS} ($= I_0/\sqrt{2}$ for a sine wave). In the U.K., the mains electricity supply is alternating, about 250 V (RMS) at a frequency of 50 hertz. In the U.S. it is 220 V (RMS) at a frequency of 60 hertz. *Compare* direct current.

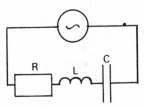

LCR circuit

L and R in series

alternating-current circuit A circuit containing a resistance R, capacitance C, and inductance L, with an alternating voltage supply, is called an *LCR circuit*. The simplest type is one in which L, C, and R are all in series. The impedance of such a circuit is given by
$$Z = \sqrt{[(X_L - X_C)^2 + R^2]}$$
where X_L is the reactance of the inductor ($2\pi fL$), and X_C is the reactance of the capacitor ($1/2\pi fC$). The current I is given by V/Z. There is a phase difference between the current in the circuit and the voltage. Current lags behind voltage by a phase angle ϕ:
$$\tan\phi = (X_L - X_C)/R$$
See also resonance.

alternator A generator for producing an alternating electric current.

AM *See* amplitude modulation.

Amagat's experiments *See* Andrews' experiments.

amalgam An alloy of mercury with one or more other metals. Amalgams may be liquid or solid.

ammeter A meter used to measure electric current. Ammeters have to have low resistance as they are connected in series in the circuit. Commonly, moving-coil instruments are used with shunt resistors to increase the current range. For alternating current a rectifier is necessary. Moving-iron instruments can be used both for d.c. and a.c. High-frequency currents may be measured with a hot-wire instrument.

amorphous Denoting a solid that has no crystalline structure; i.e. there is no long-range ordering of atoms. Many substances that appear to be amorphous are in fact composed of many tiny crystals. Soot and glass are examples of truly amorphous materials. *See also* glass.

amount of substance Symbol: n A measure of the number of entities present in a substance. *See* mole.

ampere Symbol: A The SI base unit of electric current, defined as the constant current that, maintained in two straight parallel infinite conductors of negligible circular cross section placed one metre apart in vacuum, would produce a force between the conductors of 2×10^{-7} newton per metre.

ampere balance *See* current balance.

Ampere–Laplace law *See* Ampère's law.

Ampère's law 1. (Ampère-Laplace law) The elemental force, dF, between two current elements, I_1dl_1 and I_2dl_2, parallel

5

to each other at a distance r apart in free space is given by:

$dF = \mu_0 I_1 dl_1 I_2 dl_2 \sin\theta / 4\pi r^2$

Here μ_0 is the permeability of free space and θ is the angle between either element and the line joining them.

2. The principle that the sum or integral of the magnetic flux density B times the path length along a closed path round a current-carrying conductor is proportional to the current I. For a circular path of radius r round a long straight wire in a vacuum, $B.2\pi r = \mu_0 I$. (μ_0 is the magnetic permeability of free space.) Ampère's law enables the value of B inside a solenoid to be calculated using the equation $B.dl = \mu_0 I$.

ampere-turn Symbol: At The SI unit of magnetomotive force (m.m.f.) equal to the magnetomotive force produced by a current of one ampere flowing through one turn of a conductor. See also magnetic circuit.

amplification factor See triode.

amplifier A device that increases an electrical signal applied to it as an input. If the input is an alternating voltage, the output voltage has a similar waveform with an increased amplitude.

The ratio of the output signal to the input signal (called the gain), will usually vary with the signal frequency. Amplifiers are usually designed to give a particular current, voltage, or power gain over the required frequency range. Some circuits containing a number of amplifying stages can cope with frequencies from 0 hertz (steady direct current) to radiofrequencies. In modern solid-state electronics, all of the amplifier circuit components, including many individual amplifying stages, are manufactured in a single integrated circuit.

amplitude The maximum value of a varying quantity from its mean or base value. In the case of a simple harmonic motion — a wave or vibration — it is half the maximum peak-to-peak value.

amplitude modulation (AM) A type of modulation in which the amplitude of a carrier wave is modulated by an imposed signal, usually at audio frequency.

In this way communication of a signal is made between two distant points using a radio transmission as carrier. When the carrier wave is received the audio component is extracted by the process of demodulation, and the original sound may be reproduced. See also carrier wave, demodulation.

amplitude of accommodation The eye's range of accommodation in terms of power (in dioptres). It is given by $(1/u_1 - 1/u_2)$, where u_1 is the distance from the near point to the lens and u_2 is the distance from the far point to the lens.

amu See atomic mass unit.

analyser A device for determining the plane of polarization of plane-polarized radiation. Maximum intensity is transmitted if the plane is parallel with the analyser's direction of polarization; the intensity is a minimum (theoretically zero) if the two are perpendicular. For visible radiation, analysers are usually Polaroid sheets or Nicol prisms.

anastigmatic lens A lens designed so as to minimize its astigmatic aberration. Anastigmatic lenses have different curvatures in different directions; the surface of an anastigmatic lens is part of a toroid.

AND gate See logic gate.

Andrews' experiments Experiments performed (1863) on the effect of pressure and temperature on carbon dioxide. Andrews used two thick-walled glass capillary tubes, one containing dry carbon dioxide and the other dry nitrogen. The top end of each tube was closed and the bottom end contained a plug of mercury to trap the gas. The bottom ends of the tubes were sealed into a case containing water, and pressure could be applied by means of a pair of screws. In this way Andrews achieved

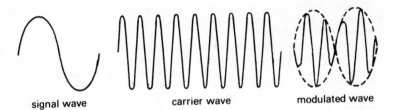

signal wave　　　　　carrier wave　　　　modulated wave

Amplitude modulation

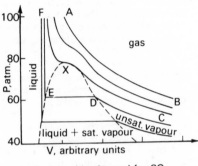

gas

Andrews' isothermal for CO$_2$

pressures up to above 10 MPa. The nitrogen was used to measure the pressure by assuming that it obeyed Boyle's law. The apparatus was surrounded by a constant temperature bath, so that isothermals (p-V curves) could be plotted at different temperatures.

In this way Andrews showed the behaviour near the critical temperature, and the liquefaction of carbon dioxide by pressure below the critical temperature. Similar experiments were done on carbon dioxide and other gases by Amagat.

anechoic chamber (dead room) A room designed so that there is little or no reflection of sound from its internal walls. The walls are covered with pyramid shapes so that stationary waves are not produced between parallel surfaces. They are coated with absorbing material. Anechoic chambers are used for experiments in acoustics.

aneroid (non-liquid) barometer See barometer.

angle of deviation See deviation.

angle of dip See inclination.

angle of incidence The angle between a ray incident on a surface and the normal to the surface at the point of incidence.

angle of polarization See Brewster angle.

angle of reflection The angle between a ray reflected by a surface and the normal to the surface at the point of reflection.

angle of refraction The angle between a ray refracted at the surface between two media and the normal to the surface at the point of refraction.

angstrom Symbol: Å A unit of length defined as 10^{-10} metre. The angstrom is sometimes used for expressing wavelengths of light or ultraviolet radiation or for the sizes of molecules.

angular acceleration Symbol: α The rotational acceleration of an object about an axis:
$$\alpha = d\omega/dt \text{ or } \alpha = d^2\theta/dt^2$$
Here ω is angular velocity; θ is angular displacement. Angular acceleration is directly analogous to linear acceleration, a. See also equations of motion, rotational motion.

angular displacement Symbol: θ The rotational displacement of an object about an axis. If the object (or a point on it) moves from point P_1 to point P_2 in a plane perpendicular to the axis, θ is the angle P_1OP_2, where O is the point at which the perpendicular plane meets the axis. *See also* rotational motion.

angular frequency (pulsatance) Symbol: ω The number of complete rotations per unit time. A simple harmonic motion of frequency f can be represented by a point moving in a circular path at constant speed. The foot of a perpendicular from the point to a diameter of the circle moves with simple harmonic motion. The angular frequency of this motion is $2\pi f$, where f is the frequency. The unit is the hertz.

angular magnification (magnifying power) Symbol: M The ratio of the angle subtended at the eye by an image to that subtended by the object: $M = \theta_1/\theta_0$. The object and image are considered to be at their actual positions, except in the case of microscopes. Here it is conventional to measure θ_0 for the object at the standard near-point distance (250 mm from the eye). The maximum useful magnifying power depends on the resolving power of the viewing system — i.e. the acuity of the eye or the grain of the photographic emulsion. *See also* magnification.

angular momentum Symbol: L The product of the moment of inertia of a body and its angular velocity. *See also* rotational motion.

angular velocity Symbol: ω The rate of change of angular displacement: $\omega = d\theta/dt$. *See also* rotational motion.

anharmonic oscillator A system whose vibration, while still periodic, cannot be described in terms of simple harmonic motions (i.e. sinusoidal motions). In such cases, the period of oscillation is not independent of the amplitude.

anion A negatively charged ion, formed by addition of electrons to atoms or molecules. In electrolysis anions are attracted to the positive electrode (the anode). *Compare* cation.

anisotropy A medium is anisotropic if a certain physical quantity differs in value in different directions. Most crystals are anisotropic electrically; important polarization properties result from differences in transmission of electromagnetic radiation in different directions. *Compare* isotropy.

annealing The process of heating a solid to a temperature below the melting point, and then cooling it slowly. Annealing removes crystal imperfections and strains in the solid.

annihilation A reaction between a particle and its antiparticle; for example, between an electron and a positron. The energy produced is equivalent to the sum of the rest masses of the annihilating particles and their kinetic energies. In order that momentum be conserved two photons are formed, moving away in opposite directions. This radiation (*annihilation radiation*) is in the gamma-ray region of the electromagnetic spectrum.
Annihilation also can occur between a nucleon and its antiparticle. In this case mesons are produced.

annual variation The direction and strength of the Earth's magnetic field at any point changes with time. This must be allowed for by navigators. One such change is a variation with a period of a year, but there are others. The amplitude of the annual variation is greatest during maximum sun-spot activity.
See also Earth's magnetism, magnetic variation.

annular eclipse *See* eclipse.

anode In electrolysis, the electrode that is at a positive potential with respect to the cathode. In any electrical system, such as a discharge tube or a solid-state electronic device, the anode is the terminal at which electrons flow out of the system.

anomalous dispersion A discontinuity in the curve of refractive constant against wavelength caused by high absorptivity of the medium at certain wavelengths. It occurs at wavelengths in the region of absorption bands in the spectrum of the absorbing substance. *See* dispersion.

anomalous expansion An increase in volume resulting from a decreased temperature. Most liquids increase in volume as their temperature rises. The density of the liquid falls with increased temperature. Water, however, shows anomalous behaviour. Between 0 and 4°C the density increases with increasing temperature.

antiferromagnetism A phenomenon found in certain solids that have two or more types of atom with magnetic moments. The magnetic moments of one type can align antiparallel with those of the other type. In antiferromagnetism the susceptibility increases with temperature up to a certain point (the *Néel temperature*). Above this temperature the material becomes paramagnetic. Ferrimagnetism is a particular form of antiferromagnetism. *See also* magnetism.

antimatter Matter formed of antiparticles. Nuclei of antimatter would consist of antiprotons and antineutrons, and would be surrounded by orbiting positrons. When matter encounters antimatter annihilation occurs.

antinodal line A line joining the antinodes (positions of maximum disturbance) in an interference pattern. *See* interference.

antinode A point of maximum vibration in a stationary wave pattern. *Compare* node.

antiparallel Having parallel lines of action that are directed in opposite directions.

antiparticle A particle of the same mass and spin, but opposite charge (and other properties) to its corresponding particle.

For example, a proton and antiproton both have mass 1836 times that of an electron and spin ½ unit, but the charge on the proton is $+1$ unit, while that on the antiproton is -1 unit. For unstable particles, such as an isolated neutron, the particle and antiparticle have the same half-life. For uncharged particles the antiparticle is indicated by a bar above the symbol, such as $\bar{n}$ for the antineutron. For charged particles the distinction is indicated by the sign, for example, e^+ is the positron, the antiparticle of an electron.

aperture A measure of the effective diameter (d) of a mirror or lens compared with its focal distance (f):
$$\text{aperture} = d/f$$
Thus a 50 mm camera lens may be used with an aperture diameter of 12.5 mm. Then, aperture $= 12.5/50$. This is usually described with the *f-number*. In this case the aperture diameter is $f/4$, often written as f4.

The transmitted light intensity depends on aperture diameter, so that I is proportional to d^2. However, large apertures lead to large aberrations although diffraction effects are more serious at small apertures. In many optical instruments, iris diaphragms vary the aperture to obtain the optimum results.

aplanatic lens A lens designed so as to minimize both its astigmatic and coma aberration.

apochromatic lens A lens designed to correct for chromatic aberration at three different wavelengths. Apochromatic lenses are constructed of three or more kinds of glass. They thus have better correction than achromatic lenses, which correct at two different wavelengths (usually in the red and blue regions of the spectrum).

apparent depth Because radiation travels at different speeds in different media, the apparent depth or thickness of a transparent sample is not the same as its real depth or thickness. The effect is

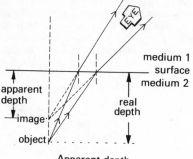

Apparent depth

very obvious when one looks down into a glass of water or a clear pool. It is associated with the fact that a long object partly submerged in water seems bent at the water surface.

The refractive constant of the substance can be defined on this basis:

refractive constant = real depth/ apparent depth

The relation is used in a number of methods for finding the refractive constant of a transparent medium. It applies to all wave radiations, not just to visible radiation.

apparent expansion *See* expansivity.

aqueous humour The watery substance between the cornea and the lens in the eye. *See* eye.

arc, electric *See* electric arc.

Archimedes' principle The upward force on an object totally or partly submerged in a fluid is equal to the weight of fluid displaced by the object. The upward force, often called *upthrust*, results from the fact that the pressure in a fluid (liquid or gas) increases with depth. If the object displaces a volume V of fluid of density ρ, then:

$$upthrust\ u = v\rho g$$

where g is the acceleration of free fall. If the upthrust on the object equals the object's weight, the object will float. *See* flotation, law of.

armature 1. The part of an electric motor or generator that carries the principal current. This is the rotating coil in a small motor but the stationary coil in a large motor or generator. Torque acting on the armature enables work to be done against the load. *See also* electric motor, rotor, stator. **2.** The moving part of any electro-mechanical device, such as an electric bell or relay.

artificial radioactivity Radioactivity induced by bombarding stable nuclei with high-energy particles, such as with neutrons in a nuclear reactor. For example:

$$^{27}_{13}Al + ^1_0n \rightarrow ^{24}_{11}Na + ^4_2He$$

represents the bombardment of aluminium with neutrons, which produces an artificially radioactive isotope of sodium of mass 24 and an alpha particle. Artificially produced radioactive nuclides can decay in a variety of ways; for example $^{24}_{11}Na$ decays by beta emission with a half-life period of about 15 h. All the transuranic elements (atomic numbers 93 and above) are artificially radioactive since they do not occur in nature.

asdic *See* sonar.

astable circuit (pulse generator) A multivibrator circuit that switches continuously and regularly from one state to another. Unlike other forms of multivibrator, no trigger pulse is needed. It is used in computers as a source of clock pulses for counting, because the output is a rectangular voltage waveform.

In the astable multivibrator, two transistors are arranged with the base terminal of each connected to the collector terminal of the other through capacitors $C1$ and $C2$ respectively. There is a steady voltage supply. $C1$ charges and $C2$ discharges until the transistors switch from one state to another and the charging direction reverses. The value of the capacitances and resistances determines the switching frequency. *See also* multivibrator.

astatic coils Two identical coils connected together in series and suspended on the same axis. When a current passes through them, any external magnetic field will result in the same turning force on each, but in opposite directions. Thus neither the Earth's magnetic field, nor any other external magnetic disturbance, will affect the rotation of the axis.

astatic pair Two identical magnetic needles suspended on the same vertical axis with their N- and S-poles pointing in opposite directions. The rotating forces on the needles from an external magnetic field, such as the Earth's, are equal and opposite. Astatic pairs are used in very sensitive galvanometers in which the current-carrying coils are wound round each needle in opposite directions. The current therefore rotates them both in the same direction and external magnetic effects are cancelled out.

astigmatism 1. A common eye defect in which the observer cannot focus on horizontal objects and vertical objects at the same distance at the same time. The cause is usually a non-spherical cornea. Visual astigmatism may be corrected with a lens with a suitable degree of cylindrical curvature. See anastigmatic lens.
2. See aberration.

astronomical unit (au, AU) The mean distance between the Sun and the Earth, used as a unit of distance in astronomy for measurements within the solar system. It is approximately 1.496×10^{11} metres.

astronomical telescope See telescope.

atmolysis The separation of a mixture of gases by using their different rates of diffusion.

atmosphere See standard pressure.

atmospheric pressure See pressure of the atmosphere.

atom The smallest part of an element that can take part in a chemical reaction. Atoms consist of a small dense positively charged nucleus, made up of neutrons and protons, with electrons in a cloud around this nucleus. The chemical reactions of an element are determined by the number of electrons (which is normally equal to the number of protons in the nucleus). All atoms of a given element have the same number of protons (the proton number). A given element may have two or more isotopes, which differ in the number of neutrons in the nucleus.

The electrons surrounding the nucleus are grouped into *shells* — i.e. main orbits around the nucleus. Within these main orbits there may be sub-shells. These correspond to atomic orbitals. An electron in an atom is specified by four quantum numbers:
(1) The *principal quantum number* (n), which specifies the main energy levels. n can have values 1, 2, etc. The corresponding shells are denoted by letters K, L, M, etc., the K shell ($n = 1$) being the nearest to the nucleus. The maximum number of electrons in a given shell is $2n^2$.
(2) The *orbital quantum number* (l), which specifies the angular momentum. For a given value of n, l can have possible values of $n-1$, $n-2$, ... 2, 1, 0. For instance, the M shell ($n = 3$) has three sub-shells with different values of l (0, 1, and 2). Sub-shells with angular momentum 0, 1, 2, and 3 are designated by letters s, p, d, and f.
(3) The *magnetic quantum number* (m) This can have values $-l$, $-(l - 1)$ ... 0 ... $+ (l + 1)$, $+ l$. It determines the orientation of the electron orbital in a magnetic field.
(4) The *spin quantum number* (m_s), which specifies the intrinsic angular momentum of the electron. It can have values $+1/2$ and $-1/2$.
Each electron in the atom has four quantum numbers and, according to the Pauli exclusion principle, no two electrons can have the same set of quantum numbers. This explains the electronic structure of atoms. See also Bohr theory.

atom bomb A bomb in which the explosion is caused by a fast uncontrolled fission reaction. *See* nuclear weapon.

atomic clock An apparatus for measuring time by the frequency of radiation emitted or absorbed in transitions of atoms. *See* caesium clock.

atomic heat *See* Dulong and Petit's law.

atomicity The number of atoms per molecule of a compound. Methane, for instance has an atomicity of five (CH_4).

atomic mass unit (amu) Symbol: u A unit of mass used for atoms and molecules, equal to 1/12 of the mass of an atom of carbon-12. It is equal to $1.660\,33 \times 10^{-27}$ kg.

atomic number *See* proton number.

atomic orbital *See* orbital.

atomic pile A nuclear reactor, particularly the early form constructed by piling up graphite blocks (the moderator) and uranium rods (the fuel).

atomic theory The theory that matter is made up of atoms that combine to form molecules. Each chemical element has a particular type of atom, which may join with like atoms to form molecules of the element, or with atoms of other elements to form molecules of a compound. †The atom consists of a dense positively charged nucleus containing protons and neutrons, surrounded by electrons. The number of protons in the nucleus determines the number and distribution of the electrons, which are held by the positive charge of the nucleus. Because the outer electrons form the chemical bonds between atoms, the chemical properties of an element depend on the electronic structure of the atom, and therefore also on the number of protons. The number of neutrons in the nucleus may vary, forming different isotopes of an element. These cannot usually be separated by chemical means.

atomic weight *See* relative atomic mass.

attenuation 1. The reduction of intensity of a radiation as it passes through a medium. It includes reductions due to both absorption and scattering. **2.** Reduction in current, voltage, or power of an electrical signal passing through a circuit.

atto- Symbol: a A prefix denoting 10^{-18}. For example, 1 attometre (am) = 10^{-18} metre (m).

AU (au) *See* astronomical unit.

audibility, limits of The frequencies beyond which sound cannot be heard by the human ear. The lowest audible frequency is about 20 hertz (a deep rumble), and the highest 15–20 kilohertz (a very high-pitched whistle). Because hearing deteriorates continuously with age, older people cannot detect sounds as high as children. *See also* infrasound, ultrasonics.

audiofrequency A frequency within the audible frequency range (about 20 hertz to about 20 kilohertz). Sound vibrations in this range can be detected by the human ear. Audiofrequency electrical signals are converted directly into sound in a loudspeaker.

audiometer A device for measuring the frequency range of the human ear and the minimum intensity of sound that can be detected at the different frequencies. It consists of a signal generator used to feed a tone of variable frequency and intensity through earphones.

avalanche A process such as that in which a single ionization leads to a large number of ions. The electrons and ions produced ionize more atoms, so that the number of ions multiplies quickly. *See* Geiger counter.

Avogadro constant Symbol: N_A The

number of particles in one mole of a substance. Its value is $6.022\,52 \times 10^{23}$.

Avogadro's law Equal volumes of all gases at the same temperature and pressure contain equal numbers of molecules. It is often called *Avogadro's hypothesis*. It is strictly true only for ideal gases.

axis *See* principal axis.

azeotropic mixture (azeotrope) A mixture of two liquids that boils without any change in composition. The proportions of components in vapour are the same as in the liquid. Azeotropic mixtures cannot be separated by distillation.

B

back e.m.f. An e.m.f. that opposes the normal flow of electric charge in a circuit or circuit element.
1. In some electrolytic cells a back e.m.f. is caused by the layer of hydrogen bubbles that builds up on the cathode as hydrogen ions pick up electrons and form gas molecules. *See also* polarization.
2. *See* self-induction.

ballistic galvanometer An instrument that measures the total electric charge passing through it in a sudden pulse of current. It is a moving-coil instrument constructed and calibrated so that the maximum deflection of the pointer is proportional to the total charge that has passed. The coil suspension is lightly damped in a ballistic galvanometer. Provided that the discharge through it occurs in a much shorter time than the suspension's natural period of oscillation, the maximum deflection is proportional to the total charge.

ballistic pendulum A device for measuring the velocity of a projectile (e.g. a bullet). It consists of a heavy pendulum, which is struck by the projectile. The velocity can be calculated by measuring the displacement of the pendulum and using the law of constant momentum.

ballistics The study of the motion of projectiles.

Balmer series A series of lines in the spectrum of radiation emitted by excited hydrogen atoms. The lines correspond to the atomic electrons falling into the second lowest energy level, emitting energy as radiation. The wavelengths (λ) of the radiation in the Balmer series are given by:
$$1/\lambda = R(1/2^2 - 1/n^2)$$
where n is an integer and R is the Rydberg constant. *See* Bohr theory. *See also* spectral series.

band, energy *See* energy bands.

band spectrum A spectrum that appears as a number of bands of emitted or absorbed radiation. Band spectra are characteristic of molecules. Often each band can be resolved into a number of closely spaced lines. The bands correspond to changes of electron orbit in the molecules. The close lines seen under higher resolution are the result of different vibrational states of the molecule.
See also spectrum.

band theory (of solids) *See* energy bands.

band width An indication of the range of frequencies, or wavelengths (wave band) that:
(1) an aerial can receive efficiently;
(2) a radio receiver or amplifier can efficiently handle;
(3) exist in a radio transmission above and below the carrier wave frequency.
See also carrier wave.

bar A unit of pressure defined as 10^5

pascals. The *millibar* (mb) is more common; it is used for measuring atmospheric pressure in meteorology.

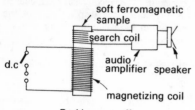

Barkhausen effect

Barkhausen effect A phenomenon that demonstrates the domain theory of magnetism. When a ferromagnetic substance is being magnetized, changes of induction occur as domains reverse direction. The effect is demonstrated as shown in the diagram; a series of clicks is heard when the current is switched on and off.

barn Symbol: b A unit of area defined as 10^{-28} square metre. The barn is sometimes used to express the effective cross-sections of atoms or nuclei in the scattering or absorption of particles.

barometer A device that measures the pressure of the atmosphere: the standard value is around 100 kPa.

The *liquid barometer* has a column of liquid in a vertical tube. Various types of *mercury barometer* are commonly used. As the external atmospheric pressure rises and falls, the length of the liquid column rises and falls.

The *aneroid* (*non-liquid*) *barometer* employs a thin-metal evacuated box. Changes in atmospheric pressure move the sides of the box, and levers communicate this movement to a pointer. In general, it is not as accurate as a liquid barometer, but it is much easier to transport and use, and is much cheaper.

The liquid barometer provides an absolute measure; aneroid barometers must be calibrated.

See also barometric height.

barometric height The 'height' of a column in a liquid barometer. As the usual barometer liquid is mercury (because of its very high density), barometric height is usually measured in millimetres of mercury (mmHg); 1 mmHg is about 135 Pa. The standard value of the pressure of the atmosphere

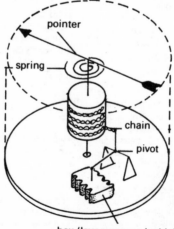

An aneroid barometer (not to scale)

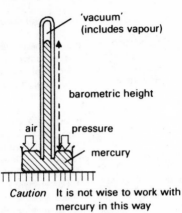

'vacuum'
(includes vapour)

barometric height

air

pressure

mercury

Caution It is not wise to work with
mercury in this way

A simple liquid barometer (not to scale)

is 760 mmHg, just over 100 kPa. *See also* STP.

baryons A group of heavy elementary particles, which includes protons and neutrons. The baryons form a subclass of the hadrons. They are further subdivided into nucleons and hyperons. *See also* elementary particles.

base *See* transistor.

base unit A unit that is defined in terms of reproducible physical phenomena or prototypes, rather than of other units. The metre, for example, is a base unit in the SI, being defined in terms of the wavelength of a particular light emission. *See also* SI units.

battery A number of similar units, such as electric cells, working together. Many dry 'batteries' used in radios, flashlights, etc., are in fact single cells. If a number of identical cells are connected in series, the total e.m.f. of the battery is the sum of the e.m.f.s of the individual cells. If the cells are in parallel, the e.m.f. of the battery is the same as that of one cell,

but the current drawn from each is less (the total current is split amongst the cells).

beam A group of rays of light, or other forms of radiation, moving in the same direction, but not necessarily parallel. Strictly, a beam is the entire set of rays coming from a point or area of an object. A *pencil* is a narrow beam from a single point.

beat frequency *See* beats.

beats A regular increase and decrease in intensity of sound waves (or other waves) caused by two waves of slightly different frequencies being added together. The waves successively reinforce and cancel each other as they move in and out of phase. Sometimes radiofrequency waves produce audiofrequency beats in sound equipment. The frequency of the resulting signal (the *beat frequency*) is given by the difference in frequencies of the two signals; i.e. $f_1 - f_2$.

If two waves of equal amplitude (a) produce beats, the resulting amplitude (A) is given by:

$$A = 2a\cos[2\pi(f_1 - f_2)t - \theta]/2$$
where θ is the phase angle between the original signals.
See also interference.

becquerel Symbol: Bq The SI unit of activity of radioactive nuclides. The activity in becquerels of a sample at a given time is the average number of disintegrations per second of its atoms at that time. *See also* curie.

bel *See* decibel.

Bernoulli effect The relation between the pressure in a steadily flowing fluid, and its velocity. The pressure is less where the velocity is higher as, for example, where water flows through a narrower section in a pipe. The pressure that lifts an aircraft also depends on this effect. If the pressure difference causing the flow is Δp, the fluid density is ρ, and the pressure and velocity at a particular point are p and v respectively, then Bernoulli's theorem states that:
$$p + \tfrac{1}{2}\rho v^2 + \Delta p = \text{constant.}$$
The equation is derived by applying the principle of conservation of energy to the kinetic energy of the flow and the work done by the pressure difference Δp.

beta decay A type of radioactive decay in which a nucleus emits, for instance, an electron. The result is a nuclide with the same mass number but a proton number one greater (electron emission) than the original nuclide. An example of beta decay is:
$$^1_1H \rightarrow {}^3_2He + e^- + \nu$$
The particles emitted in beta decay are *beta particles*. Streams of beta particles are *beta rays* or *beta radiation*. The particles are about 100 times more penetrating than alpha particles of the same energy.
Beta particles may have a range of energies up to a maximum value characteristic of the nucleus concerned. The total energy is constant; it is carried by the beta particle and an antineutrino emitted at the same time. In another type of beta decay, positrons are emitted.

In such cases the excess energy is carried by a neutrino. An example is:
$$^{13}_7N \rightarrow {}^{13}_6C + e^+ + \nu$$
See also alpha decay.

beta transformation The transformation of a nucleus by beta decay. Also the decay of a neutron to a proton, an electron, and an antineutrino:
$$n \rightarrow p + e^- + \nu$$

betatron A device for accelerating electrons to very high energies (300 MeV or more). Electrons produced from a source are injected into an evacuated doughnut-shaped ring between the poles of an electromagnet just as the magnetic field is being increased. As the magnetic field increases the electrons are accelerated, making as many as a quarter of a million circuits before the magnetic field reaches its maximum, at which time the orbit is changed by passing a current through auxiliary coils to deflect the electrons onto a target. A betatron can be compared to a transformer in which a cloud of electrons in the toroid constitutes the secondary circuit. Alternating current circulates in the primary coil (the magnetizing coil) but the electrons are extracted at the end of a quarter cycle before the decreasing primary current can cause deceleration.

BeV *See* GeV.

bevatron The name given to the proton synchrotron at the University of California, which can accelerate protons to energies of 10^{28} joules (6 GeV).

bias A potential applied to an electrode in an electronic device to produce the desired characteristic.

biaxial crystal A type of birefringent crystal having two axes parallel to which the ordinary ray and the extraordinary ray travel at the same speed.

biconcave Describing a lens with two concave faces. *Compare* biconvex. *See also* lens.

biconvex Describing a lens with two convex faces. *Compare* biconcave. *See also* lens.

bimetallic strip A strip of two metals joined side by side. When heated, the metals expand by different amounts, causing the strip to bend. Bimetallic strips are used in thermostats and safety cut-outs.

Calculation of mass defect and binding energy

mass of neutrons (4 × 1.008 983 amu)	4.035 932 amu
mass of protons (3 × 1.008 144 amu)	3.024 432 amu
total mass of constituents	7.060 364 amu
mass of ^{7}Li nucleus	7.018 222 amu
mass defect	0.042 142 amu
binding energy (1 amu = 931.14 MeV)	39.240 MeV

binding energy (of a nucleus) The energy equivalent to the difference between the mass of the nucleus and the sum of the masses of its constituent nucleons. An example of calculating the binding energy of $^7_3$Li, with 4 neutrons and 3 protons is shown.
A useful measure is binding energy per nucleon. In the example the binding energy per nucleon is 39.2501/7 = approximately 5.6 MeV. For most nuclei, binding energy lies between about 7 and about 9 MeV per nucleon, reaching a maximum of about 9 MeV for nuclei of mass number about 60. The difference in mass in the example (i.e. the mass equivalent to the binding energy) is the *mass defect*.

binoculars An optical instrument providing a telescope for each eye, thus giving distance perception as well as magnification.
Prism binoculars use a pair of prisms inside each telescope. These reflect rays by internal reflection. Their effect is to bring the inverted image upright, reduce the telescope length, and allow the object lenses to be further apart than the eyes (thus improving stereoscopic vision). Binoculars are often described thus: 15 × 40. The first figure is the magnification; the second is the aperture of each object lens.
Opera glasses are a simpler low-power device, consisting of two Galilean telescopes side by side. The telescopes produce upright images without the need for extra inverting lenses or prisms.

binocular vision Vision using two eyes. The brain forms a single three-dimensional view from the two separate images. This type of vision (*stereoscopic vision*) gives more information about distance and shape than monocular vision could.

bioluminescence *See* luminescence.

Biot–Savart law The elemental field strength dB at a point distant r from a current element Idl in free space is given by:
$$dB = \mu_0 Idl\sin\theta/4\pi r^2$$

bipolar transistor *See* field effect transistor.

biprism, Fresnel's A glass prism with a large angle, used to produce two coherent (virtual) sources for light interference experiments. As with Young's double slit arrangement, the wavelength λ of the incident monochromatic radiation is given by:
$$\lambda = yd/D$$
where y is the fringe separation, d is the source separation, and D is the source-screen distance. The fringes obtained with this arrangement are brighter than those in Young's experiment.

birefringent crystal A crystal that splits incident transmitted light into two beams, each polarized perpendicularly to the other. The effect (called *birefringence* or *double refraction*) is particularly well-known in calcite (Iceland spar). It depends on the angle of incidence relative to the crystal axes, along which the speed of the light differs.

17

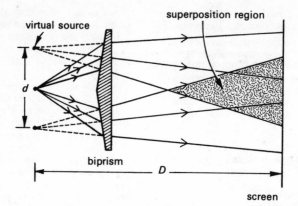

Biprism, Fresnel's

The ordinary ray obeys the laws of refraction: it is polarized perpendicularly to the crystal axis. The extraordinary ray does not obey the laws of refraction (in the usual sense); hence its name. The study of the polarization properties of crystals is of great significance in geology, for the identification of minerals.

bistable circuit (flip-flop) An electronic circuit, usually a multivibrator, that has two stable states and is switched from one to the other by a trigger pulse. Bistable circuits are used in computer logic for counting and storing binary digits (0 and 1). They form the basis of several different logic gates.

In a bistable multivibrator the input pulse is fed to the base terminal of one transistor (*TR1*) through a resistor *R1* and directly to the collector of the other transistor (*TR2*). The base of *TR2* and the collector of *TR1* are connected through a resistor *R2*. In logic circuits, bistable circuits may have two or perhaps more inputs. These are connected so that the output level (high or low) depends on whether one or both inputs are high. Sometimes a square-wave input is used as a clock or counter. *See also* logic gate, multivibrator.

Bitter pattern A microscopic pattern of domain boundaries on a ferromag-

netic surface, made visible by painting the surface with a colloidal suspension of very small iron particles. The production of Bitter patterns is similar to showing the shape of a magnet under a sheet of paper by using iron dust.

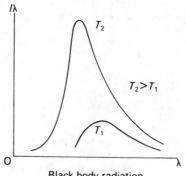

Black body radiation

black body A body that absorbs all the radiation falling on it. The absorptance and emissivity of a black body are both equal to 1. In practice, a small hole in a uniform-temperature enclosure acts as a black body.

The radiation from a black body covers the whole wavelength range (sometimes the alternative term *full radiator* is used). The distribution of power with wavelength of this *black-body radiation* has a characteristic form. As the temperature increases the amount of radiation increases and the maximum in the curve moves to longer wavelengths. A black body radiates mainly infrared radiation below about 800 K. Visible radiation does not predominate until the temperature is above about 6000 K.

blanket A layer of fertile material surrounding the core in a breeder reactor.

blind spot The area of the retina where the optic nerve leaves the eye. It is not sensitive to light as in this region there is no layer of rods and cones. See also eye.

Bloch wall The boundary of a magnetic domain, over the width of which the atomic magnetic moment directions change. In iron, Bloch walls are around 100 nm thick.

block and tackle See pulley.

blooming A method of coating lenses to reduce back-reflection from their surfaces. It involves destructive interference in the thin layer. Each such layer can completely prevent reflection at only one wavelength (λ). λ is four times the layer thickness (t). For best effects the coating medium used should have a refractive constant $n = \sqrt{(n_1 n_2)}$, where n_1 and n_2 are the refractive constants of the media on each side. Single-layer blooming is normally used to prevent reflection of yellow light; bloomed surfaces thus reflect reds and blues, and appear purple. Multilayer blooming is sometimes employed, but is very costly.

blue shift See red shift.

Board of Trade unit (BTU) A unit of energy equivalent to the kilowatt-hour (3.6×10^6 joules). It was formerly used in the U.K. for the sale of electricity.

Bohr theory A theory introduced by Niels Bohr (1911) to explain the spec-

trum of atomic hydrogen. The model he used was that of a nucleus with charge $+e$ orbited by an electron with charge $-e$ moving in a circle radius r. If v is the velocity of the electron, the centripetal force, mv^2/r is equal to the force of electrostatic attraction $e^2/4\pi\epsilon_0 r^2$. Using this, it can be shown that the total energy of the electron (kinetic and potential) is $-e^2/8\pi\epsilon_0 r$.

If the electron is considered to have wave properties, then there must be a whole number of wavelengths around the orbit, otherwise the wave would be a progressive wave. For this to occur $n\lambda = 2\pi r$, where n is an integer, 1, 2, 3, 4, ... The wavelength, λ, is h/mv, where h is the Planck constant and mv the momentum. Thus for a given orbit:

$$nh/2\pi = mvr$$

This means that orbits are possible only when the angular momentum (mvr) is an integral number of units of $h/2\pi$. Angular momentum is thus quantized. In fact, Bohr in his theory did not use the wave behaviour of the electron to derive this relationship. He assumed from the beginning that angular momentum was quantized in this way. Using the above expressions it can be shown that the electron energy is given by

$$E = -me^4/8\epsilon_0^2 n^2 h^2$$

Different values of n (1, 2, 3, etc.) correspond to different orbits with different energies; n is the principal quantum number. In making a transition from an orbit n_1 to another orbit n_2 the energy difference ΔW is given by:

$$\Delta W = W_1 - W_2 = me^4(1/n_2^2 - 1/n_1^2)/8\epsilon_0^2 h^2$$

This is equal to $h\nu$ where ν is the frequency of radiation emitted or absorbed. Since $\nu\lambda = e$, then

$$1/\lambda = me^4(1/n_1^2 - 1/n_2^2)/8\epsilon_0^2 ch^3$$

The theory is in good agreement with experiment in predicting the wavelengths of lines in the hydrogen spectrum, although it is less successful for larger atoms. Different values of n_1 and n_2 correspond to different spectral series, with lines given by the expression:

$$1/\lambda = R(1/n_1^2 - 1/n_2^2)$$

R is the *Rydberg constant*. Its experimental value is $1.096\,78 \times 10^7$ m^{-1}. The value from Bohr theory $(me^4/8\pi\epsilon_1^2ch^2)$ is 1.097 00 $\times$ 10^7 m^{-1}. *See also* atom.

boiling A change from liquid to gas occurring at a characteristic temperature (the *boiling point*). Boiling occurs when the saturated vapour pressure of the liquid equals the external pressure. Bubbles of vapour can then form in the liquid. The temperature at which this happens must depend on external pressure; boiling points are therefore usually quoted at standard pressure. *See also* change of state, latent heat.

boiling-water reactor (BWR) A nuclear reactor in which water (in contact with the fuel elements) is used as both coolant and moderator.

Boltzmann constant Symbol: k The constant 1.380 54 J K^{-1}, equal to the gas constant (R) divided by the Avogadro constant (N_A). *See also* degrees of freedom.

bomb calorimeter A sealed insulated container, used for measuring energy released during combustion of substances (e.g. foods and fuels). A known amount of the substance is ignited inside the calorimeter in an atmosphere of pure oxygen, and undergoes complete combustion at constant volume. The resultant rise in temperature is related to the energy released by the reaction. Such energy values (*calorific values*) are often quoted in joules per kilogram (J kg^{-1}).

boron chamber A device for detecting low-energy neutrons, in which a compound of boron (usually the gas BF$_3$) fills an ionization chamber. The ^{10}B nuclei, which constitute 18% of natural boron, absorb neutrons and emit alpha particles, which are detected by the ionization they cause.

boson *See* elementary particles.

boundary layer A thin layer of fluid, such as the one next to a solid surface past which the fluid is moving. Friction with the surface slows flow within the boundary layer so that next to the surface the fluid is stationary. At the other edge of the boundary layer, the velocity approaches that of the main flow. Within it the effects of viscosity are significant, whereas in the main stream they can often be neglected.

Boyle's law At a constant temperature, the pressure of a fixed mass of a gas is inversely proportional to its volume: i.e. $pV = K$, where K is a constant. (A graph of p against $1/V$ is a straight line.) The value of K depends on the temperature and on the gas. The law holds strictly only for ideal gases. Real gases follow Boyle's law at low pressures and high temperatures. *See* gas laws.

Boyle temperature *See* gas laws.

Boys' experiment A method of determining the gravitational constant, G. A short beam (about 25 cm) with gold spheres at each end was suspended horizontally by a quartz torsion fibre. Measurements were made of the period of the torsional oscillations of the beam and of its angular deflection when large dense masses were placed near each sphere. In this way, G could be calculated. The method was more accurate than Cavendish's experiment for G.

Bragg's law If a beam of X-rays of wavelength λ is directed at a crystal with parallel crystal planes that are distance d apart, then the reflected X-rays from each plane undergo interference. Constructive interference occurs at angles θ where $n\lambda = 2d\sin\theta$, n being an integer (1, 2, 3, etc.). θ is the angle between the crystal plane and the incident beam (called the *Bragg angle*). The equation is used in determining crystal structure from interference patterns produced by monochromatic X-rays.

breeder reactor A nuclear reactor that produces additional nuclei at a rate greater than that at which fuel is consumed. The core fuel consists of a fissile element, for example, uranium enriched to about 25% in the ^{235}U isotope. The core is surrounded by a blanket of fertile material, mostly ^{238}U in the form of natural or depleted uranium. Some of the surplus neutrons from the fission of ^{235}U convert ^{238}U into ^{239}Pu, which is fissile. A primary circuit of liquid sodium can be used through the core to carry heat away. The heat is transferred to a secondary circuit of liquid sodium that boils water, the steam then operating turbines and generators as in a conventional power station. Such a reactor is also termed a *fast breeder reactor* or *fast reactor* because the neutrons moving through the core and blanket are fast moving, being of high energy (several MeV) as compared to those in thermal reactors (about 0.025 eV). The possibility of ^{232}Th as a fertile material has not yet been exploited. *See also* nuclear reactor.

bremsstrahlung X-rays emitted when fast electrons are slowed down violently, as when electrons strike the target in an X-ray tube. The word translates as 'braking radiation'. Bremsstrahlung is caused when an electron passes through the electric field of a nucleus and constitutes the continuous X-ray spectrum.

Brewster angle Symbol: i_B The angle of incidence, on a partially reflecting surface, at which the reflected radiation is fully plane-polarized. It is also the angle of incidence at which the reflected and refracted beams are perpendicular. Polarization by reflection is a refractive property of the surface.
$$_1n_2 = \tan i_B$$
The plane of polarization is parallel to the surface. The refracted radiation is partly polarized parallel to the normal. Formerly, the Brewster angle was called the *angle of polarization* or the *polarizing angle.*

bridge circuit An arrangement of four electrical components in a square with an input across two opposite corners

and an output across the other opposite corners. *See* Wheatstone bridge.

brightness A vague term describing the intensity of light. It can be applied to a source of light, to light itself, or to an illuminated surface. The brightness or intensity of light, in any of these three cases, relates to the rate of supply of energy (i.e. the power). The relation is complicated as it must take account of the sensitivity of the eye (or other detector) at different frequencies.
See also photometry.

Brinell test A way of measuring the hardness of a material. A standard steel ball of known hardness is pressed into the material's surface with a known force. The size of the indentation indicates the hardness.

British thermal unit (Btu) A unit of energy equal to $1.055\,06 \times 10^3$ joules. It was formerly defined by the heat needed to raise the temperature of one pound of air-free water by one degree Fahrenheit at standard pressure. Slightly different versions of the unit were in use depending on the temperatures between which the degree rise was measured.

Brownian movement The random motion of small particles in a fluid — for example, smoke particles in air. The particles, which may be large enough to be visible with a microscope, are continuously bombarded by the invisibly small molecules of the fluid. *See also* kinetic theory.

brush An electrical contact with a moving part, as on an electric motor or generator.

brush discharge A form of bright gas discharge occurring near sharp points of high potential.
The potential difference causing such a discharge is lower than that necessary for a spark or arc. The discharge is characterized by luminous streamers, which take on a treelike form.

BTU *See* Board of Trade unit.

Btu *See* British thermal unit.

bubble chamber A container of a liquid kept slightly above its boiling temperature by increased pressure and used to show tracks of ionizing radiation. The liquid is often liquid hydrogen. Just before the passage of a particle the pressure is momentarily reduced, and a photograph taken. Ions formed along the paths of charged particles or gamma-ray photons act as nuclei on which bubbles form. Magnetic fields can be applied causing curvature of the paths of charged particles. Bubble chambers are more useful than cloud chambers. The greater density of the liquid increases the chance that a nuclear reaction will occur.
See also cloud chamber.

bulk modulus *See* elastic modulus.

bulk strain *See* strain.

bumping Violent boiling of a liquid caused when bubbles form at pressure above atmospheric pressure.

Bunsen burner A gas burner consisting of a vertical metal tube with an adjustable air-inlet hole at the bottom. Gas is allowed into the bottom of the tube and the gas–air mixture is burnt at the top. With too little air the flame is yellow and sooty. Correctly adjusted, the burner gives a flame with a pale blue inner cone of incompletely burnt gas, and an almost invisible outer flame where the gas is fully oxidized and reaches a temperature of about 1500°C.

Bunsen cell A type of primary cell in which the positive electrode is formed by carbon plates in nitric acid solution and the negative electrode consists of zinc plates in sulphuric acid solution.

buoyancy The tendency of an object to float. The term is sometimes used for the upward force (upthrust) on a body.
See centre of buoyancy.

buzzer A device in which a vibrating reed, operated by an electric current, produces a continuous buzzing sound.
See also electric bell.

BWR *See* boiling-water reactor.

bypass capacitor A capacitor that allows alternating currents to pass through it rather than through another component connected in parallel. The capacitance of the capacitor determines the frequency range of alternating current that can pass most easily.

C

cable, coaxial An electrical cable consisting of a central wire in an insulating sheath surrounded by a woven mesh of conductor, all inside an outer insulating sheath. Coaxial cables are not affected by external electric or magnetic disturbances or signals. They are used in transmitting high-frequency signals; for example, in television aerial connections. Normally the outer conductor is earthed.

cadmium cell *See* Weston cadmium cell.

caesium clock An apparatus used to produce the steady frequency used in defining the second. It depends on the fact that, in a magnetic field, caesium-133 atoms can have two different energy levels between which transitions occur by absorption of radiofrequency radiation of frequency 9 192 631 770 hertz. In a caesium clock, the number of atoms in the higher state is detected, and the signal used to stabilize the oscillator producing the radiation.

calcite (Iceland spar) A naturally occurring form of calcium carbonate. It is the best-known example of a mineral that shows double refraction (bire-fringence).

calendar year *See* year.

Callendar and Barnes' apparatus An apparatus for measuring the specific thermal capacity of a fluid under steady-state conditions. Essentially, it consists of a long tube with a spiral electrical heating wire down its axis. The fluid passes through the tube at a constant rate; the equilibrium input and output temperatures are measured. The tube carrying the liquid is surrounded by an evacuated jacket to minimize energy leakage. Under steady conditions:
$$VI = mc(\theta_2 - \theta_1) + w$$
I and *V* are the current and voltage in the heating wire, *m* is the mass of fluid passed in unit time, *c* the specific thermal capacity, and θ_1 and θ_2 the input and output temperatures respectively. The energy leakage, *w*, may be eliminated by taking a second set of readings using the same temperature extremes.

calomel electrode A half cell having a mercury electrode coated with mercury(I) chloride, in an electrolyte consisting of potassium chloride and (saturated) mercury chloride solution. Its standard electrode potential against the hydrogen electrode is accurately known and it is a convenient secondary standard.

caloric theory An obsolete theory that heat was a weightless fluid. The theory was abandoned in the nineteenth century.

calorie Symbol: cal A unit of energy approximately equal to 4.2 joules. It was formerly defined as the energy needed to raise the temperature of one gram of water by one degree Celsius. Because the specific thermal capacity of water changes with temperature, this definition is not precise. The mean or thermochemical calorie (cal_{TH}) is defined as 4.184 joules. The international table calorie (cal_{IT}) is defined as 4.186 8 joules.

Formerly the mean calorie was defined as one hundredth of the heat needed to raise one gram of water from 0°C to 100°C, and the 15°C calorie as the heat needed to raise it from 14.5°C to 15.5°C.

calorimeter A device for measuring thermal energy; for example in determining thermal capacity, specific latent thermal capacity, energies of combustion, etc.

camera An optical instrument able to record an image formed by visible light (or by other electromagnetic radiation). The record may be in the form of chemical changes in a photographic emulsion, or electrical signals as in a T.V. camera. A lens or system of lenses is used to focus the radiation onto the sensitive surface. Adjustment can be made for different object distances. The shutter allows light in for a set time (except in a video camera); the diaphragm varies the aperture.

camera, pin-hole *See* pin-hole camera.

Canada balsam A transparent adhesive with a refractive constant of 1.53. As this constant is similar to that of many optical media, Canada balsam has various uses in optical instrumentation. *See*, for example, Nicol prism.

canal rays Streams of positive ions obtained from a discharge tube by boring small holes in the cathode. The ions being attracted to the cathode can thus pass through it forming positive rays.

candela Symbol: cd The SI base unit of luminous intensity, defined as the intensity (in the perpendicular direction) of the black-body radiation from a surface of 1/600 000 square metre at the temperature of freezing platinum and at a pressure of 101 325 pascals.

candle power A former name for luminous intensity measured in terms of the international candle.

candle-power An outdated measure of

luminous intensity. The unit until 1948 was the international standard candle. This was first defined, in Britain, in the 1860 Gas Act; it was the brightness of a candle made in a certain way and burning at a certain rate.

capacitance Symbol: C The ability of a system of conductors and insulators to store electric charge. If, in a given case, a charge Q is maintained by a potential difference V, then the capacitance is defined as the quotient Q/V. The unit of capacitance is the farad (F). See also capacitor.

capacitor (formerly, condenser) A device for storing electric charge. It usually consists of two parallel conductors separated by some insulating material (the dielectric). The capacitance of a capacitor increases:
(1) the greater the common area of the conductors;
(2) the smaller the distance between them;
(3) the higher the relative permittivity of the dielectric.
The charge in a capacitor is stored partly by polarization of the particles of the dielectric. The capacitance (C) of a parallel-plate capacitor is given by
$$C = \epsilon A/d$$
where ϵ is the permittivity of the material between the plates, A their common area, and d their separation.
An isolated sphere has capacitance:
$$C = 4\pi\epsilon_r\epsilon_0 r$$
where r is the radius of the sphere, and $\epsilon_r\epsilon_0$ the permittivity of the medium.

capillary action The effect that occurs when a fine tube, or capillary, is placed vertically with one end in a liquid. The height of the liquid inside the tube will be above or below the level of the liquid by an amount depending on the cohesion (surface tension) of the liquid and the adhesion between liquid and tube. The narrower the capillary is, the greater the difference in height. Because water 'sticks' to glass — its adhesion tends to be greater than its cohesion — it has a concave meniscus in a glass tube, and

will be drawn up the tube by capillary action. Mercury, on the other hand, will be lowered inside a glass capillary, because here the cohesion is greater than the adhesion.
The distance h that a liquid of density ρ and surface tension γ will rise up a tube of radius r is given by the equation
$$h = 2\gamma\cos\alpha/rg\rho$$
where α is the contact angle between the meniscus and the tube. α is measured below the meniscus, so that for a convex meniscus, α is greater than 90° and h is negative.

capillary tube A tube with a narrow bore (internal diameter). Capillary tubing usually has relatively thick walls. Glass capillaries are used in mercury thermometers and experiments on surface tension. See also capillary action.

capture The absorption of one particle by another. For instance, a positive ion may capture an electron to form a neutral atom. In some nuclear reactions, an atomic nucleus may capture a neutron, with subsequent emission of a gamma-ray photon.

carbon black (soot) A form of amorphous carbon produced by incomplete combustion of gas (or other organic matter). It is used in experiments as a coating for surfaces that need to be good absorbers of radiation, for example in detectors of thermal radiation, such as the thermopile. It is also used to reduce the amount of thermal radiation emitted by a surface.

carbon cycle (carbon–nitrogen cycle) A series of nuclear reactions postulated to account for energy production in stars. In this series $^{12}_{6}C$ is an intermediary in the process by which hydrogen nuclei fuse to form helium with release of energy. The first step is the fusion of carbon and hydrogen nuclei:
$^{12}_{6}C + ^{1}_{1}H \rightarrow ^{13}_{7}N$ + gamma radiation
followed by:
$^{13}_{7}N \rightarrow ^{13}_{6}C$ + positron
$^{13}_{6}C + ^{1}_{1}H \rightarrow ^{14}_{7}N$ + gamma radiation
$^{14}_{7}N + ^{1}_{1}H \rightarrow ^{15}_{8}O$ + gamma radiation

24

$^{15}_8O \rightarrow \, ^{15}_7N$ + positron

$^{15}_7N + \, ^1_1H \rightarrow \, ^{12}_6C + \, ^4_2He$

The net result is:-

$4 \times \, ^1_1H \rightarrow \, ^4_2He$ + 2 positrons and binding energy. *See also* proton–proton chain reaction.

carbon dating (radiocarbon dating) A method of dating — measuring the age of (usually archaeological) materials that contain matter of living origin. It is based on the fact that ^{14}C, a beta emitter of half-life approximately 5730 years, is being formed continuously in the atmosphere as a result of cosmic-ray action. The ^{14}C becomes incorporated into living organisms. After death of the organism the amount of radioactive carbon decreases exponentially by radioactive decay. The ratio of ^{12}C to ^{14}C is thus a measure of the time elapsed since the death of the organic material.

Uncertainties arise because of uncertainty as to the past rate of production of ^{14}C, the possibility of exchange of carbon with carbon of a different age during the elapsed time, the possibility of contamination of the sample, and the effect of burning of fossil fuels on the composition of atmospheric carbon.

The method is most valuable for specimens of up to 20 000 years old, though it has been modified to measure ages up to 70 000 years. For ages of up to about 8000 years the carbon time scale has been calibrated by dendrochronology; i.e. by measuring the $^{12}C:^{14}C$ ratio in tree rings of known antiquity.

carbon microphone *See* microphone.

carbon–nitrogen cycle *See* carbon cycle.

Carnot cycle The idealized reversible cycle of four operations occurring in a perfect heat engine. These are the successive adiabatic compression, isothermal expansion, adiabatic expansion, and isothermal compression of the working substance. The cycle returns to its initial pressure, volume, and temperature, and transfers energy to or from mechanical work. The efficiency of the

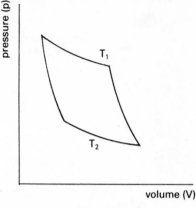

Carnot's cycle

Carnot cycle is the maximum attainable in a heat engine. *Compare* Otto cycle. *See* Carnot's principle.

Carnot's principle (Carnot theorem) The efficiency of any heat engine cannot be greater than that of a reversible heat engine operating over the same temperature range. It follows directly from the second law of thermodynamics, and means that all reversible heat engines have the same efficiency, independent of the working substance. If heat is absorbed at temperature T_1 and given out at T_2, then the Carnot efficiency is $(T_1 - T_2)/T_1$.

carrier 1. *See* carrier wave.

2. A charge carrier — either an electron or a positive hole.

3. The agent that carries the electric charge in a current, e.g. an electron, positive hole, or ion.

carrier wave An electromagnetic wave, usually in the long-wave radio to radar frequency range, that is used to carry information in a radio transmission. The information is superimposed on the carrier wave by modulation. *See also* modulation.

cascade liquefier A device for liquefying a series of gases in stages. The

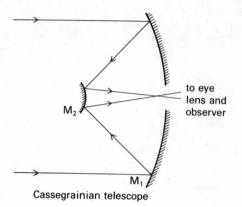

Cassegrainian telescope

critical temperature of each gas is different. A gas with a high critical temperature is liquefied by compression and evaporated under a lower pressure, cooling a second gas below its critical temperature. The second gas can also be liquefied and evaporated to produce a still lower temperature, and so on. *See also* Linde process.

cascade process Any process that takes place in a number of stages.
An example is the diffusion process used to enrich uranium. The gas, uranium hexafluoride, diffuses through a long series of porous barriers. A very slight enrichment takes place at each, because of the very slightly higher speed of diffusion of $^{235}UF_6$ compared with that of $^{238}UF_6$. The gas-centrifuge enrichment process is also a cascade process.

Cassegrainian telescope A type of reflecting telescope in which the converging mirror has a hole in its centre. Light is reflected back onto a diverging mirror, then forward through the hole into the eyepiece. *See also* reflector.

cathetometer An optical instrument for accurately measuring heights or lengths; a small telescope or microscope mounted so that it can be slid along a graduated scale.

cathode In electrolysis, the electrode that is at a negative potential with respect to the anode. In any electrical system, such as a discharge tube or a solid-state electronic device, the cathode is the terminal at which electrons enter the system.

cathode rays Streams of electrons given off by the cathode of a gas-discharge tube at low pressure.

cathode-ray tube (CRT) An electron tube that converts electrical signals into a pattern on a screen. The cathode-ray tube forms the basis of the cathode-ray oscilloscope and the television receiver. It consists of an electron gun, which produces an electron beam. The electrons are focused onto and moved across a luminescent screen by magnetic and/ or electric fields, to give a small moving spot of light. *See also* cathode-ray oscilloscope.

cathode-ray oscilloscope (CRO) An instrument for investigating electrical signals. An electron beam is directed at a cathode-ray tube forming a small spot.

26

The beam can be deflected in the horizontal and vertical directions by an electrostatic field from two sets of plates. The horizontal deflection is often a constant repeating sweep of the spot produced by an internal time base. The signal to be investigated is amplified and applied to the vertical deflecting plates. Thus, a graph of the signal is 'drawn' on the screen.

cation A positively charged ion, formed by removal of electrons from atoms or molecules. In electrolysis, cations are attracted to the negatively charged electrode (the cathode). *Compare* anion.

caustic The curve (surface in three dimensions) on which rays meeting a curved reflector or lens meet after reflection or refraction. *See* focal point.

Cavendish's experiment A method of determining the gravitational constant, *G*. A beam with small lead spheres at each end was suspended horizontally by a torsion wire. Large spheres of lead were then placed close to the smaller spheres on opposite sites of the beam in order to turn the beam through an angle. The lead spheres were then moved to the opposite sides of the beam, turning the beam in the opposite direction. *G* could be calculated from the total angular deflection. The method, which was the first attempt to measure *G*, was less accurate than Boys' experiment.

cell 1. A system having two plates (electrodes) in a conducting liquid (electrolyte). An *electrolytic cell* is used for producing a chemical reaction by passing a current through the electrolyte (i.e. by electrolysis). A *voltaic* (or *galvanic*) cell produces an e.m.f. by chemical reactions at each electrode. Electrons are transferred to or from the electrodes, giving each a net charge. *See also* accumulator, Daniell cell, electrolysis, Leclanché cell.
2. *See* photocell, photoelectric cell.
3. *See* Kerr effect.

Celsius scale A temperature scale in which the temperature of melting pure ice is taken as 0° and the temperature of boiling water 100° (both at standard pressure). The *degree Celsius* (°C) is equal to the kelvin. This was known as the *centigrade scale* until 1948, when the present name became official. Celsius' original scale was inverted (i.e. had 0° as the steam temperature and 100° as the ice temperature).
See also temperature scale.

centi- Symbol: c A prefix denoting 10^{-2}. For example, 1 centimetre (cm) = 10^{-2} metre (m).

centigrade scale *See* Celsius scale.

centimetric waves Electromagnetic waves with wavelengths in the range 1-10 cm (i.e. part of the microwave region).

central force A force that acts on any affected object along a line to an origin. Electric forces between charged particles are central; frictional forces are not.

centre of buoyancy (for an object in a fluid) The centre of mass of the displaced fluid volume. For a floating object to be stable, the centre of mass of the object must lie below the centre of buoyancy; when the object is in equilibrium, the two lie on a vertical line. *See also* Archimedes' principle.

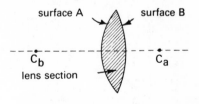

Centre of curvature

centre of curvature Each surface of a simple curved lens or mirror is part of a sphere. The centre of curvature c of such a surface is the centre of the sphere

27

of which the surface is part. *See also* lens, mirror.

centre of gravity *See* centre of mass.

centre of mass A point in a body (or system) at which the whole mass of the body may be considered to act. Often the term *centre of gravity* is used. This is, strictly, not the same, except if the body is in a constant gravitational field. The centre of gravity is the point at which the weight may be considered to act.

The centre of mass coincides with the centre of symmetry if the body has a uniform density throughout. In other cases the principle of moments may be used to locate the point. The *centroid* of a plane or solid shape is the point at which the centre of mass would be if the shape or solid were of uniform-density material.

centre of pressure If a surface lies horizontally in a fluid, the pressure at all points will be the same. The resultant force will then act through the centroid of the surface. If the surface is not horizontal, the pressure on it will vary with depth. The resultant force will now act through a different point; the centre of pressure is not at the centroid. The centre of pressure of a submerged surface

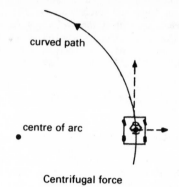

curved path

centre of arc

Centrifugal force

is the point through which the resultant of the pressure forces acts.

centrifugal force A force supposed to act radially outwards on a body moving in a curve. In fact there is no real force acting; centrifugal force is said to be a 'fictitious' force, and the use is best avoided. The idea arises from the effect of inertia on an object moving in a curve. If a car is moving around a bend, for instance, it is forced in a curved path by friction between the wheels and the road. Without this friction (directed towards the centre of the curve) the car would continue in a straight line. The driver also moves in the curve,

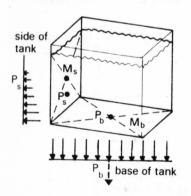

side of tank

P_s

M_s

P_s

P_b M_b

P_b base of tank

P_b = centre of pressure of base
M_b = centre of mass of base
$P_b = M_b$
P_s = centre of pressure of side
M_s = centre of mass of side in contact with liquid

Centre of pressure

constrained by friction with the seat, restraint from a seat belt, or a 'push' from the door. To the driver it appears that there is a force radially outwards pushing his body out — the centrifugal force. In fact this is not the case; if the driver fell out of the car he would move straight forward at a tangent to the curve. It is sometimes said that the centrifugal force is a 'reaction' to the centripetal force — this is *not* true. (The 'reaction' to the centripetal force is an outward push on the road surface.) *See also* centripetal force.

centrifuge A device for separating the components of a suspension by spinning. In simple types, the liquid containing suspended particles is placed in a tube, which is whirled round in a horizontal circle. Particles less dense than the liquid move to the top of the tube; denser particles accumulate at the bottom.

The device thus accelerates the process of separation that would result from gravitational forces. The gravitational pressure gradient is replaced by a much stronger pressure gradient related to centripetal forces. An *ultracentrifuge* is a high-speed centrifuge suitable for separating very small particles or even large molecules. Gas centrifuges are used in isotope separation.

centripetal force A force that causes an object to move in a curved path rather than continuing in a straight line. The force is provided by, for instance:
— the tension of the string, for an object whirled on the end of a string;
— gravity, for an object in orbit round a planet;
— electric force, for an electron in the shell of an atom.

The centripetal force for an object of mass m with velocity v, and path radius r is mv^2/r, or $m\omega^2 r$, where ω is angular velocity. A body moving in a curved path with a velocity v has an acceleration because the direction of the velocity changes, even though the magnitude of the velocity may remain constant. This acceleration, which is directed towards

the centre of the curve, is the *centripetal acceleration*. It is given by v^2/r or $\omega^2 r$.

centroid *See* centre of mass.

c.g.s. system A system of units that uses the centimetre, the gram, and the second as the base mechanical units. Much early scientific work used this system, but it has now almost been abandoned. In the c.g.s. system there are two sets of electrical units: *electrostatic units* in which the permittivity of free space (ϵ_0) is unity; and *electromagnetic units* in which the permeability of free space (μ_0) is unity.

chain reaction The progressive disintegration of fissile material (e.g. ^{235}U) by bombardment with neutrons, which in turn results in the production of more neutrons. These may, under suitable conditions, produce further fissions. Fission of ^{235}U yields a varying number of neutrons depending on the energy of the incident neutrons. Neutrons may escape from the mass of uranium or be absorbed by nonfissile nuclei and so be ineffective in promoting the chain reaction. If one fission produces neutrons which, in turn, cause more than one fission, there is a branching chain reaction, which may be explosive. On the other hand, if insufficient neutrons are captured the chain reaction will die out. To sustain a chain reaction in uranium the proportion of the fissile isotope ^{235}U has to be increased by *enrichment* and the fast neutrons have to be slowed down, usually by the presence of a moderator so that they can be absorbed by ^{235}U nuclei. The artificial radioactive isotope ^{239}Pu can also sustain a chain reaction.

change of state A change from one state of matter to another. *See also* latent heat.

characteristic In an electronic device, such as a diode or a transistor, a graph that shows how a voltage or current between two of the terminals varies with

Changes of state

solid – liquid	melting (fusion)
liquid – solid	freezing
liquid – gas	evaporation
	(boiling at boiling temperature)
gas – liquid	condensation
solid – gas	sublimation
gas – solid	condensation

respect to voltage or current at other terminals of the device.

charge (electric charge) Symbol: Q A basic property of some elementary particles of matter. Charge has no definition; rather it is taken as a basic experimental quantity. There are two types of charge conventionally called *positive* and *negative*. Like charges repel each other and unlike charges attract each other. The unit of charge is the coulomb (C). The charge on a body arises from an excess or deficit of negative electrons with respect to positive protons.

charcoal An amorphous form of carbon made by heating wood or other organic material in the absence of air. *Activated charcoal* is charcoal heated to drive off absorbed gas. It is used for absorbing gases and for removing impurities from liquids.

charge density The electric charge per unit volume of a material per unit area of a surface or per unit length of a line. *Volume charge density* (Symbol: ρ) is measured in coulombs per cubic metre ($C\,m^{-3}$). *Surface charge density* (Symbol: σ) is measured in coulombs per square metre ($C\,m^{-2}$). The unit of *linear charge density*, λ, is the coulomb per metre ($C\,m^{-1}$). For a conductor, excess charge is distributed over the surface; the surface charge density increases with curvature.

Charles' law For a given mass of gas at constant pressure, the volume increases by a constant fraction of the volume at 0°C for each Celsius degree rise in temperature. The constant fraction (α) has almost the same value for all gases — about 1/273 — and Charles' law can be written in the form
$$V = V_0(1 + \alpha_v\theta)$$
where V is the volume at temperature θ°C and V_0 the volume at 0°C. The constant α_v is the thermal expansivity of the gas. For an ideal gas its value is 1/273.15.

A similar relationship exists for the pressure of a gas heated at constant volume:
$$p = p_0(1 + \alpha_p\theta)$$
Here, α_p is the pressure coefficient. For an ideal gas $\alpha_p = \alpha_v$, although they differ slightly for real gases. It follows from Charles' law that for a gas heated at constant pressure, $V/T = K$, where T is the thermodynamic temperature and K is a constant. Similarly, at constant volume p/T is a constant.

Charles' volume law is sometimes called *Gay-Lussac's law* after its independent discoverer. *See also* absolute temperature, gas laws.

chemical hygrometer *See* hygrometer.

chemiluminescence *See* luminescence.

chemisorption *See* adsorption.

chirality *See* optical activity.

choke An inductor used to reduce the high-frequency components of an alternating signal by presenting a higher impedance for them. Chokes are also used for smoothing fluctuations in the output current of a rectifier circuit.

choroid The middle of the three layers of the eye. *See* eye.

chromatic aberration *See* aberration.

circuit, electrical A combination of electrical components that form a conducting path. When the path is broken, for example by a switch, the circuit is said to be *open*. When a complete loop is formed allowing current to flow as a result of a potential difference in the circuit, it is said to be *closed*. A *short circuit* is one in which there is a low resistance path through which the current can flow. *See also* integrated circuit.

circuit element A resistor, capacitor, inductor, transistor, or other device used in making up electric circuits.

circular measure Measurement of angles in radians.

circular motion A form of periodic (or cyclic) motion, that of an object moving at constant speed v in a circle of constant radius r. For this to be possible, a positive central force must act. The terms used to describe circular motion are shown in the table. *See also* centripetal force, rotational motion.

circular polarization A type of polarization of electromagnetic radiation in which the plane of polarization rotates uniformly round the axis as the ray progresses. Circularly polarized light is equivalent to (and is produced by) the combination of two equal plane-polarized rays moving together but out of phase by 90°.

Clark cell A type of cell formerly used as a standard source of e.m.f. It consists of a mercury cathode coated with mercury sulphate, and a zinc anode. The electrolyte is zinc sulphate solution. The e.m.f. produced is 1.434 5 volts at 15°C.

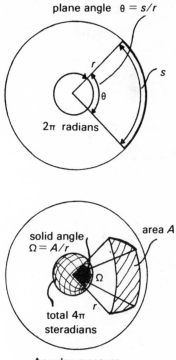

plane angle $\theta = s/r$

2π radians

solid angle $\Omega = A/r$

area A

total 4π steradians

Angular measure

The Clark cell has been superseded as a standard by the Weston cadmium cell.

classical physics The part of physics that was developed before, and therefore does not include, either quantum theory or the theory of relativity. For instance, Newton's laws of motion belong to classical mechanics; they are a special case of the theory of relativity applying

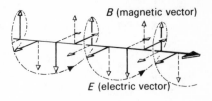

B (magnetic vector)

E (electric vector)

Circular polarization

31

classical physics

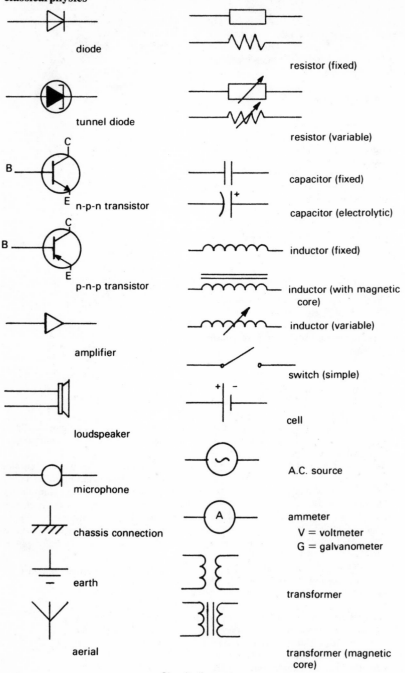

diode

resistor (fixed)

tunnel diode

resistor (variable)

capacitor (fixed)

n-p-n transistor

capacitor (electrolytic)

B

inductor (fixed)

inductor (with magnetic core)

p-n-p transistor

inductor (variable)

amplifier

switch (simple)

cell

loudspeaker

A.C. source

microphone

ammeter
V = voltmeter
G = galvanometer

chassis connection

earth

transformer

aerial

transformer (magnetic core)

Circuit elements

32

Quantities used in describing circular motion

Quantity	Symbol	Formula
angular displacement	θ	–
angular velocity	ω	$\omega = v/r$
centripetal acceleration	a	$a = v^2/r = \omega^2 r$
centripetal force	F	$F = mv^2/r = m\omega^2 r$
frequency	f	$f = \omega/2\pi$
period	T	$T = 1/f$

at velocities that are small in comparison with the speed of light.

cleavage The splitting of a crystal along planes of atoms, to form smooth surfaces.

clinical thermometer A thermometer designed for taking body ('blood') temperature; a mercury thermometer with a kink in the thread just above the bulb, and calibrated in the range 95-115°F. The mercury reaches a maximum value but does not then return to the bulb until shaken down past the kink.

closed circuit See circuit, electrical.

closed system (isolated system) A set of one or more objects that may interact with each other, but do not interact with the world outside the system. This means that there is no net force from outside or energy transfer. Because of this the system's angular momentum, energy, mass, and linear momentum remain constant.

cloud chamber A chamber used to show the tracks of ionizing radiation, especially alpha and beta particles.
A *diffusion cloud chamber* has felt strips near the top soaked in water and ethanol. The bottom of the chamber is held at a low temperature, and there is continuous diffusion of vapour down the chamber. At one particular level, water droplets condense only along the tracks of ionizing radiation. The *expansion cloud chamber* contains moist air, sometimes

with ethanol vapour, which is cooled by a sudden adiabatic expansion causing the air to become supersaturated with water vapour. Water droplets condense out preferentially on ions formed along the tracks.
It can be arranged that passage of an ionizing particle through the chamber is detected by a counter and the resulting pulse can be used to operate the pump, so that the expansion takes place just after the ion pairs have been formed. A camera may also be triggered to take a photograph of the tracks just as they become visible. Magnetic fields can be applied and the resulting curvature of the tracks provides information about the charge and energy of the particles. *See also* bubble chamber.

coaxial Denoting bodies or shapes that have a common axis of rotation or radial symmetry.

Cockcroft-Walton accelerator The first particle accelerator (1931), used to achieve the first artificial nuclear disintegration by bombarding a lithium target with protons (to produce alpha particles). The device was a simple form of linear accelerator; potential differences up to 800 kV were achieved by a voltage multiplier device.

coefficient of expansion See expansivity.

coefficient of friction See friction.

coercive force The magnetizing field intensity required to reduce the flux

33

density in a material to zero. If the material is magnetically saturated the coercive force is a maximum; it is then referred to as the *coercivity*. *See also* hysteresis cycle.

coercivity *See* hysteresis cycle.

coherent Describes waves, or sources, that are always in phase — that is, the peaks and troughs always come together. The laser is a single source of visible coherent radiation. At longer wavelengths it becomes easier to produce coherent waves.

coherent units A system or sub-set of units (e.g. SI units) in which the derived units are obtained by multiplying or dividing together base units, with no numerical factor involved.

cohesion *See* surface tension.

coincidence circuit A device that detects the occurrence of two events that either coincide or occur within a specified time interval of each other. Signals recording the two events are fed into the device, which produces an output only if the two impulses occur within the specified time. Such a device is the basis of a coincidence counter.

cold emission Emission of electrons from a solid by a process other than thermionic emission. The term is usually used to describe either field emission or secondary emission.

collector *See* transistor.

collimator An arrangement for producing a parallel beam of radiation for use in a spectrometer or other instrument. A system of lenses and slits is utilized.

colloid A substance containing very small particles (sizes in the range 10^{-9}–10^{-5} m). Sols, gels, and emulsions are examples of colloids.

colorimeter An instrument able to give readings for any colour in terms of the intensity of each of the three primary colours.

colour A physiological sensation produced in the eye-brain system relating to the wavelength of visible radiation. Traditionally the visible spectrum is divided into seven colour regions. However, individual eyes differ in their sensitivities to different wavelengths and the spectral colours also merge gradually into each other. In the retina of the eye, certain cells (cones) are responsible for colour vision; they cannot operate at low light levels, which is why colour vision is not so effective at night.

Monochromatic radiations of different wavelengths are called *hues*; the normal human eye can distinguish yellow hues only 1 nm apart. If a hue is mixed with white light, it is an 'unsaturated' hue, now called a *tint*. Different saturated hues mix together, either to produce the effect of other hues, or of nonspectral colours. The study of colour mixing by addition of coloured lights (hues) shows that any colour sensation can be produced by mixing three primary hues. It is normal to choose a red, a green, and a blue for this purpose. Coloured television pictures are produced on this basis.

While 'white' light is defined as the normal mixture of all visible radiations, the same sensation can be obtained by mixing complementary pairs of lights (hues).

The apparent colours of surfaces in white light depends on the wavelengths that are reflected. Thus yellow paint (pigment) reflects mainly yellow wavelengths (with perhaps some orange and green); the rest are absorbed. When two paints of different colours are mixed, the colour reflected will be the sum of the wavelengths absorbed by neither pigment. This is 'colour mixing by subtraction'. The three usual primary colours are the greenish-blue 'cyan' (white light with the red absorbed); the crimson magenta (green absorbed), and yellow (blue absorbed).

colour blindness *See* colour vision.

colour vision The ability of the eye to distinguish different colours. The mechanism is not fully understood. It is thought that there may be three types of cone cell in the retina, able to react to red, green, and blue light. Nerve signals corresponding to the intensities of each of these give colour impressions in the brain.

There are various forms of *colour blindness*: they differ in rarity. True colour blindness — the lack of any colour vision — is very rare. Most commonly people (especially males) have difficulties with reds, greens, browns, and pale tints. *See also* colour.

coma *See* aberration.

common emitter *See* transistor.

commutator The part of a direct-current electric motor or generator that connects the coil(s) to the outside circuit, changing the connections round as the coil rotates. *See* electric motor.

comparator An apparatus used in measurements of the expansivity of solids. A horizontal bar of the material is used with a fine scratch near each end. The bar is surrounded by a water bath and two vertical microscopes are used, focused on the scratches on each end. Changes of length, produced by changes in temperature, are measured by the microscopes: a standard bar at constant temperature is used for comparison.

compass A device used to show magnetic force field direction. It consists of a small magnet pivoted so that it is free to move in a horizontal plane. In the Earth's magnetic field a compass points along the magnetic meridian. A scale, graduated in degrees, may be placed underneath the magnet to assist in navigation. *See also* magnetometer.

complementary colours Two coloured lights (hues) that mix to produce the sensation of white. There is an infinite number of pairs of complementary colours. The spectral hues can be plotted around a roughly triangular 'chromaticity curve' that has white at the centre. The members of each pair of complementary colours lie at opposite sides of white on this curve.
See also colour.

component forces *See* component vectors.

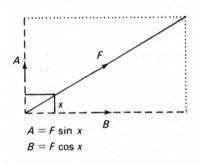

$A = F \sin x$
$B = F \cos x$

Component vectors

component vectors The components of a given vector (such as a force or velocity) are two or more vectors with the same effect as the given vector. In other words the given vector is the resultant of the components. Any vector has an infinite number of sets of components. Some sets are more use than others in a given case, especially pairs at 90°.

component velocities *See* component vectors.

compound lens Two or more lenses used together as a unit. For instance, the eyepiece of a telescope and the lens of a camera are both normally compound lenses; each has a number of elements (the single lenses) along the same optical axis. Insect compound eyes are not compound lenses in this sense — the elements are not on a single axis.

The elements of a compound lens may or may not touch. The function of compound (rather than single) lenses is usually to minimize aberrations. *See also* doublet.

compound microscope *See* microscope, compound.

compound pendulum *See* pendulum.

compressibility The reciprocal of the bulk modulus of a material.

Compton effect An increase in the wavelength of X-rays or gamma rays when scattered by free electrons. In practice, it is observed with solids in which the valence electrons are 'nearly free'. It cannot be explained by treating the electromagnetic radiation as waves; instead it must be regarded as composed of particles — photons — which collide with the electrons. In such a collision energy is transferred to the electron. The photon energy ($h\nu$) decreases; consequently the frequency (ν) of the radiation decreases and the wavelength increases.

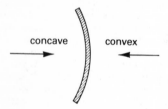

Concave and convex surfaces

concave Curving inward, away from the viewpoint. A *concave mirror* is one with a concave reflecting surface. A *concave lens* is either a biconcave or a plano-concave lens. *See also* convex, lens, mirror.

concavo-convex Describes a lens with one concave surface and one convex surface. Most spectacle lenses have this shape. Concavo-convex lenses can be converging or diverging. They are sometimes called *meniscus lenses. See also* lens.

condensation A change from vapour to liquid — the reverse of evaporation. *See also* dew, relative humidity.

condenser 1. An early term for capacitor. 2. A lens, set of lenses, or mirror in an optical instrument used to concentrate light onto the object to be viewed. *See* microscope (compound), projectors.

condenser microphone A type of microphone in which the diaphragm is one plate of a charged capacitor ('condenser'). Sound waves move the diaphragm, thus altering the capacitance. The corresponding changes in potential difference generate the signal.

conduction band *See* energy bands.

conduction 1. A process of net energy transfer through a substance without movement of the substance itself. The rate of transfer depends on the sample length and cross-sectional area, the temperature difference, and the nature of the material.
   Good conductors, such as copper and silver, are often metals that have free electrons. The energy is transmitted by movement of these. In poor solid conductors (insulators) there are no conduction electrons and the transfer is only by vibrations of atoms or molecules in the crystal structure.
See also conductivity, thermal.
2. Passage of charge through a sample under the influence of an electric field. The charge may be carried by electrons (e.g. in metals), by electrons and positive holes (in electrolytes), or by ions (in electrolytes and gases). *See also* energy bands.

conductivity, electrical Symbol: σ The ability of a material to conduct electric current; the reciprocal of the resistivity. Conductivity does not depend on the dimensions of the conductor.
   It is also the ratio of current density to

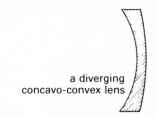

a converging
concavo-convex lens

a diverging
concavo-convex lens

Types of concavo-convex lens

electric field strength. The unit is the siemens per metre ($S\ m^{-1}$).

conductivity, thermal Symbol: λ, k A measure of the ability of a material to conduct energy. It is defined as the energy transfer per unit time per unit cross-sectional area of the conductor, per unit temperature gradient along the conductor. The unit of thermal conductivity is the watt metre^{-1} kelvin^{-1} (W $m^{-1}\ K^{-1}$). For a parallel-sided block of material with no loss through the sides:
$$E/t = \lambda A(\theta_2 - \theta_1)/l$$
where E is the amount of energy transferred in a time t, A the cross-sectional area, l the length, and θ_2 and θ_1 the temperatures of the faces. In general:
$$dE/dt = -\lambda A d\theta/dt$$
See also Lees' disc, Searle's bar.

conductor, electrical A substance with a relatively high electrical conductivity. Many metals are good electrical conductors because of their free electrons. Graphite is also a good conductor. Many other non-metals tend to be poor conductors. See also energy bands.

conductor, thermal A substance that has a relatively high thermal conductivity. In general, many metals are good conductors of energy — copper and silver are notable examples. Many non-metallic solids are poor conductors (i.e. they are *thermal insulators*). Gases are particularly poor conductors (or good insulators).

cones The cells in the retina of the eye that are important for vision in normal light. They are also responsible for colour vision. Their mechanism of action is not fully understood. See also eye.

conjugate points Any pair of points relative to a lens, reflector, or other optical system such that either is imaged at the other. Compare the symmetry in object distance u and image distance v in the equation:
$$1/v + 1/u = 1/f$$

conservation law A physical law that some property of a system remains constant throughout a series of changes. Examples are the laws of conservation of mass, energy, momentum, and charge.

conservation of angular momentum, law of See constant angular momentum, law of.

conservation of charge The principle that the total net charge of any closed system is constant.

conservation of energy, law of See constant energy, law of.

conservation of (linear) momentum,

law of *See* constant (linear) momentum, law of.

conservation of mass, law of *See* constant mass, law of.

conservation of momentum, law of *See* constant (linear) momentum, law of.

conservation of mass and energy The law that the total energy (rest mass energy + kinetic energy + potential energy) of a closed system is constant. In most chemical and physical interactions the mass change is undetectably small, so that the measurable rest-mass energy does not change (it is regarded as 'passive'). The law then becomes the classical *law of conservation of energy*; the inclusion of mass in the calculation is necessary only in the case of nuclear changes or systems involving very high velocities. *See also* mass–energy equation, rest mass.

conservative force A force such that, if it moves an object between two points, the energy transfer (work done) does not depend on the path between the points. It must then be true that if a conservative force moves an object in a closed path (back to the starting point), the energy transfer is zero. Gravitation is a conservative force; friction is not.

constant A quantity or factor that remains at the same value when others are changing. *See also* fundamental constants.

constantan An alloy of copper (50-60%) and nickel used in thermocouples and precision resistors. Its electrical resistance varies very little with changes in temperature.

constant angular momentum, law of (law of conservation of angular momentum) The total angular momentum of a system cannot change unless a net outside torque acts. *See also* constant (linear) momentum, law of.

constant energy, law of (law of conservation of energy) The total energy of a system cannot change unless energy is taken from or given to the outside. The law is equivalent to the first law of thermodynamics. *See also* mass–energy equation.

constant mass, law of (law of conservation of mass) The total mass of a system cannot change unless mass is taken from or given to the outside. *See also* mass–energy equation.

constant (linear) momentum, law of (law of conservation of (linear) momentum) The total *linear momentum* of a system cannot change unless a net outside force acts.

constant momentum, law of *See* constant (linear) momentum, law of.

constant pressure gas thermometer *See* gas thermometer.

constant volume gas thermometer *See* gas thermometer.

constructive interference *See* interference.

contact angle An angle formed between solid and liquid surfaces. For example, when a drop of mercury rests (in air) on top of a horizontal glass surface, the contact angle measured inside the drop from the glass to the mercury-air interface, is greater than 90°. This is because the mercury does not wet the glass — its cohesion is greater than the adhesion to glass. Water does wet the surface of clean glass and the contact angle is less than 90°. *See also* capillary action.

contact potential A potential difference produced between two solids that are in contact. The variation of contact potential with temperature is the basis of the thermocouple.

containment 1. The process of containing radioactive material within a shield for safety reasons.

2. The process of containing the plasma in a fusion reactor so that it does not come in contact with the walls of the vessel. Magnetic fields are used.

continuous spectrum A spectrum composed of a continuous range of emitted or absorbed radiation. Continuous spectra are produced in the infrared and visible regions by hot solids. *See* black body. *See also* spectrum.

continuous wave An electromagnetic wave that is transmitted continuously over a period of time, as in radio communication. It is distinguished from a pulsed or intermittent wave as used in radar. *See also* radar.

control grid *See* grid.

control rods Rods of neutron-absorbing material that can be moved into or out of the core of a nuclear reactor to control the chain reaction. Control rods are made of boron, cadmium, or other substance that absorbs neutrons.

convection The transfer of energy by flow of a liquid or gas. In *natural convection* the fluid flow is caused by temperature differences between one part of the fluid and another. For example, in an electric kettle the heating element raises the temperature of the water next to it, which expands and rises. Colder water then flows in beneath it, setting up a convection current. In *forced convection*, energy is carried away from the source by flow produced by a pump or fan.

In natural convection the rate of loss of energy from a body to the surroundings is proportional to $\theta^5/4$ where θ is the excess temperature over the surroundings. In forced convection it is proportional to θ. *See also* Newton's law of cooling.

conventional current An electric current considered to flow from points at positive potential to points at negative potential. In typical conductors, charge is carried by electrons, which flow in the

opposite direction. In semiconductors, the charge may be carried by positive holes, flowing in the same direction as the conventional current.

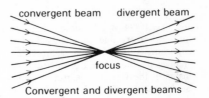
Convergent and divergent beams

convergent Coming together. A convergent beam becomes narrower as it travels. Its narrowest point is called the *focus*; after passing through the focus, the beam will be divergent (moving apart).

converging lens A lens that can refract a parallel beam into a convergent beam. Converging lenses used in air are thicker in the middle than at the edge. They may be biconvex, plano-convex, or concavo-convex in shape. As these lenses have positive power, they are sometimes called *positive lenses*. *Compare* diverging lens. *See also* lens.

converging mirror (converging reflector) A mirror that can reflect a parallel beam into a convergent beam. Converging reflectors always have concave surfaces. The section shape is the arc of a circle in simple cases; the arc of a parabola is needed for more precise work. As these mirrors have positive power, they are sometimes called *positive mirrors*. *Compare* diverging mirror. *See also* mirror.

converter A device for converting alternating current to direct current or vice versa.

converter reactor *See* nuclear reactor.

convex A *convex mirror* is one with a convex reflecting surface. A *convex lens* is either a biconvex lens or a plano-convex lens. *See also* concave, lens, mirror.

coolant A fluid that removes energy from a source (e.g. a hot engine) by convection.

cooling correction A method of correcting for errors due to energy loss in experiments that involve temperature changes. In the method of mixtures, for instance, the final temperature reached by the mixture may be slightly less than the theoretical one as a result of cooling during the change. A graph of temperature against time throughout the course of the experiment (i.e. a *cooling curve*) allows a correction to be made based on Newton's law of cooling.

coplanar forces Forces in a single plane. If only two forces act through a point, they must be coplanar. So too are two parallel forces. However, nonparallel forces that do not act through a point cannot be coplanar. Three or more nonparallel forces acting through a point may not be coplanar.

core 1. A piece of iron or other magnetic material used to concentrate the magnetic lines of force in a transformer, electromagnet, or similar device. **2.** The central part of a nuclear reactor in which the chain reaction occurs.

Coriolis force A 'fictitious' force used to describe the motion of an object in a rotating system. For instance, air moving from north to south over the surface of the Earth would, to an observer outside the Earth, be moving in a straight line. To an observer on the Earth the path would appear to be curved, as the Earth rotates. Such systems can be described by introducing a tangential Coriolis 'force'.

cornea The transparent part of the eye. *See* eye.

corona discharge Small regions of glowing air around a charged conductor at a critical potential. It is often accompanied by hissing sounds and is caused by the ionization of the air around the conductor. Corona discharges are produced at sharp points, where the surface charge density, and electric field, are high.

corpuscular theory The theory that light travels in the form of particles ('corpuscles'). The theory went out of favour in the early nineteenth century when it seemed proved that electromagnetic radiation consisted of waves. However, near the end of that century it was shown that in certain circumstances, light must be thought of as particles rather than as waves. Modern quantum theory claims that light travels in 'packets' (called quanta, or photons), which can show either wave properties or particle properties. *See* wave theory of light.

cosmic radiation High-energy radiation reaching the Earth from space. Cosmic rays are mainly protons with some electrons, alpha particles, other atomic nuclei, and gamma rays. Collisions in the upper atmosphere with nuclei produce other particles (including mesons). Gamma radiation produces electrons and positrons by pair production. These can undergo other reactions and in certain cases a single primary particle can give rise to a large shower of secondary radiation.

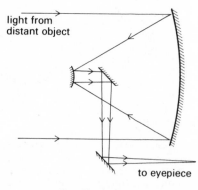

light from distant object

to eyepiece

Coudé telescope

Coudé telescope A type of reflecting telescope. Light from the concave mirror is reflected back onto a convex mirror, then onto a plane at an angle to the axis, and into the eyepiece. *See also* reflector.

coulomb Symbol: C The SI unit of electric charge, equal to the charge transported by an electric current of one ampere flowing for one second. 1 C = 1 A s.

Coulomb's law The electric force between two charges is inversely proportional to the square of their distance apart. The full law giving the mutual force between two charges Q_1 and Q_2 is
$$F = Q_1 Q_2 / 4\epsilon_r\epsilon_0 r^2$$
where r is the distance apart and $\epsilon_r\epsilon_0$ the permittivity of the medium between the charges.

coulo(mb)meter *See* voltameter.

counter A device for detecting and counting particles and photons. *See* Geiger counter, scintillation counter, spark counter, proportional counter, semiconductor counter.

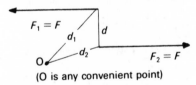

(O is any convenient point)

Couple

couple A pair of equal parallel forces in opposite directions and not acting through a single point. Their linear resultant is zero, but there is a net turning-effect (moment). The resultant turning effect T (torque) is given by
$$T = d_1 F_1 + d_2 F_2$$
$$T = (d_1 - d_2)F$$
$$T = dF$$

creep A slow deformation of a solid as a result of a continuous applied stress.

critical angle *See* total internal reflection.

critical damping *See* damping.

critical density *See* critical state.

critical mass The minimum mass of material that can sustain a nuclear chain reaction. An explosive chain reaction (as in a nuclear fission weapon) results when subcritical masses are brought together to form a mass greater than the critical mass. In a nuclear reactor, the chain reaction is controlled by a moderator.
  The critical mass depends on the purity of the material used, the geometry of the arrangement, and the presence or absence of reflectors of neutrons around the fissile material.
See also chain reaction.

critical pressure *See* critical state.

critical specific volume *See* critical state.

critical speed In fluid flow, the speed at which the behaviour of the fluid switches from that of laminar flow to that of turbulent flow or vice versa. *See also* Reynolds number.

critical state (critical point) The state of a fluid under conditions in which the liquid and gas phases have the same density. The temperature, pressure, and density concerned are the *critical temperature*, *critical pressure*, and *critical density* respectively. The specific volume is the *critical specific volume*. Above its critical temperature a gas cannot be liquefied by increasing the pressure. The critical point corresponds to a point of inflection on an isothermal for the gas. *See* Andrews' experiments.

critical temperature *See* critical state.

CRO *See* cathode-ray oscilloscope.

cross product *See* vector product.

cross section In a collision process — for example, the bombardment of nuclei by neutrons — the apparent area that particles present to the bombarding particles. This is not its 'true' cross-sectional area, but depends on the probability of a reaction occurring. In particular, it varies with the energy of the incident particles. The measurement is used for other types of collision reactions besides neutron absorption, including reactions of atoms, ions, molecules, electrons, etc. The unit is the square metre (m^2).

crown glass *See* optical glass.

CRT *See* cathode-ray tube.

cryogenics The physics of very low temperatures, including techniques for producing low temperatures and the study of how materials behave when cooled.

cryogenics The production of low temperatures and the study of low-temperature phenomena.

cryometer A thermometer designed to measure very low temperatures.

cryoscopy The determination of freezing points.

cryostat A container for keeping samples at a constant low temperature.

crystal A solid with a regular geometric shape. In crystals the particles (atoms, ions, or molecules) have a regular three-dimensional repeating arrangement in space. This is called the *crystal structure*. The crystal *lattice* is the arrangement of points in space at which the particles are positioned. The external appearance of the crystal is the *crystal habit*. The external shape of the crystal depends on the internal arrangement of the lattice. It also depends on the way in which the crystal is formed.
See also crystalline.

crystal habit *See* crystal.

crystalline Describing a solid that has a crystal structure; i.e. a regular internal arrangement of atoms, ions, or molecules. Note that a substance can be crystalline without necessarily having a geometrical external shape.

crystallography The study of the formation, structure, and properties of crystals. *See also* X-ray crystallography.

crystal microphone *See* microphone.

crystal oscillator A fixed-frequency oscillator in which the output is derived from a crystal of Rochelle salt or quartz specially cut so that its natural frequency of mechanical vibration is the desired value. This is usually in the range of about one kilohertz to several megahertz. The action is based on the piezoelectric effect. A potential difference is applied to two electrodes attached to two parallel cut faces of the crystal. The electric field sets up a mechanical stress which causes the crystal to resonate at its natural frequency. This in turn generates an alternating potential difference across the crystal. *See also* oscillator.

crystal rectifier A semiconductor diode. *See* diode, rectifier, semiconductor.

crystal structure *See* crystal.

cubic expansivity *See* expansivity.

curie Symbol Ci A unit of radioactivity, equivalent to the amount of a given radioactive substance that produces 3.7 $\times 10^{10}$ disintegrations per second.

Curie point *See* Curie temperature.

Curie's law The susceptibility, χ, of some paramagnetic substances is inversely proportional to the thermodynamic temperature, T:
$\chi = c/T$, where c is a constant.
More generally, the relationship is that of the *Curie–Weiss law*:
$$\chi = c/(T - \theta)$$
where θ is the *Weiss constant*, a fixed

temperature characteristic of the material.

Curie temperature (Curie point) The temperature above which a ferromagnetic substance loses its ferromagnetism.

It then shows paramagnetism only; the alignment of magnetic moments in domains is overcome by thermal vibrations. The Curie temperature for iron is 760°C; that for nickel is 356°C. Metals that are paramagnetic at room temperatures could become ferromagnetic if cooled. For example, the Curie temperature of dysprosium is -188°C.

Curie-Weiss law *See* Curie's law.

current Symbol: *I* A flow of electric charge. Current is the result of motion of electrons or ions under the influence of an e.m.f. It is measured in amperes (A).

current balance (ampere balance) A device for measuring electric current to a high degree of accuracy, by measuring the electromagnetic force produced between two conductors carrying the current.

current density Symbol: *j* The electric current per unit area of conductor cross-section. It is measured in ampere metre^{-2} (A m^{-2}). In a conductor it is equal to nvQ, where n is the number of free charges per unit volume, v is their average drift velocity, and Q is their charge.

cycle A series of events that is regularly repeated (e.g. a single orbit, rotation, vibration, oscillation, or wave). A cycle is a complete single set of changes, starting from one point and returning to the same point in the same way.

cyclic accelerator *See* accelerator.

cyclotron A device for accelerating charged particles to high energies. Protons can be given energies of over 10 MeV, deuterons over 20 MeV, and alpha particles about 40 MeV. Particles are injected near the centre of an evacuated space between two D-shaped boxes placed between the poles of a strong permanent magnet. Within each 'dee' the particles describe a semicircular orbit. An alternating voltage is applied to the dees at such a frequency that the particles are accelerated by the potential difference each time they reach the gap, causing them to increase their path radii and speeds in steps. Eventually, after several thousand revolutions, they reach the perimeter of the dees at high speed, where a deflecting electric field directs them onto the target.

The energies that can be achieved are limited by the relativistic increase in mass of the particles. As their velocity approaches that of light, the increase in mass causes the period of rotation to increase so that they no longer reach the gaps in phase with the potential difference. *See also* synchrocyclotron.

D

Dalton's law (of partial pressures) The pressure of a mixture of gases is the sum of the partial pressures of each individual constituent. (The *partial pressure* of a gas in a mixture is the pressure that it would exert if it alone were present in the container.) Dalton's law is strictly true only for ideal gases.

damped oscillation An oscillation with an amplitude that progressively decreases with time. *See* damping.

damping The reduction in amplitude of a vibration with time by some form of resistance. A swinging pendulum will at last come to rest; a plucked string will not vibrate for long — in both cases internal and/or external resistive forces progressively reduce the amplitude and bring the system to equilibrium.

In most cases the damping force(s) will

be proportional to the object's speed. In any event, energy must be transferred from the vibrating system to overcome the resistance.

Where damping is an asset (as in bringing a meter pointer to rest), the optimum situation occurs when the motion comes to zero in the shortest time possible, without vibration. This is *critical damping*. If the resistive force is such that the time taken is longer than this, *overdamping* occurs. Conversely, *underdamping* involves a longer time with several vibrations of decreasing amplitude.

Daniell cell A type of primary cell consisting of two electrodes in different electrolytes separated by a porous partition. The positive electrode is copper immersed in copper(II) sulphate solution. The negative electrode is zinc-mercury amalgam in either dilute sulphuric acid or zinc sulphate solution. With sulphuric acid the e.m.f. is about 1.08 volts; with zinc sulphate it is about 1.11 volts. The porous partition allows ions to move through while the reaction is taking place, and the cell must be dismantled when not in use.

At the copper electrode copper ions in solution gain electrons from the metal and are deposited as copper atoms:
$$Cu^{2+} + 2e^- \rightarrow Cu$$
The copper electrode thus gains a positive charge. At the zinc electrode, zinc atoms from the electrode lose electrons and dissolve into solution as zinc ions, leaving a net negative charge on the electrode:
$$Zn \rightarrow 2e^- + Zn^{2+}$$

dating Methods of determining the age of archaeological or geological specimens, the Earth, meteorites, etc. The main methods of special interest in physics involve measurements of radioactivity. It is assumed that some radioactive nuclide is disappearing by decay and a product is formed. Measurements of the amount of parent nuclide, or product nuclide, or the ratio parent to product can give an estimate of the age in certain circumstances. *See* carbon

dating, fission-track dating, potassium-argon dating, rubidium–strontium dating, thermoluminescent dating, uranium–lead dating.

daughter A given nuclide produced by radioactive decay from another nuclide (the *parent*).

day The time taken for the Earth to make one complete rotation on its axis.
There are various methods of measuring this. The *solar day* is the time between two successive transits of the Sun across the meridian. The *mean solar day* is the average of this over one year. The *sidereal day* is measured with reference to a star. Its value is 23 hours 56.06 minutes.

d.c. *See* direct current.

deadbeat Describing an instrument (e.g. a galvanometer) that is damped so that its oscillations die away very quickly.

dead room *See* anechoic chamber.

de Broglie wave A wave associated with a particle, such as an electron or proton. In 1924, Louis de Broglie suggested that, since electromagnetic waves can be described as particles (photons), particles of matter could also have wave properties. The wavelength (λ) has the same relationship to momentum (p) as in electromagnetic radiation: $\lambda = h/p$, where h is the Planck constant. *See also* quantum theory.

debye Symbol: D A unit of electric dipole moment equal to $3.335\,64 \times 10^{-30}$ coulomb metre. It is used in expressing the dipole moments of molecules.

deca- Symbol: da A prefix denoting 10. For example, 1 decametre (dam) = 10 metres (m).

decay The spontaneous disintegration of unstable (radioactive) nuclei to give other nuclei, accompanied by the emission of particles and/or photons. The product of a given decay may be stable or may itself be radioactive. In a sample of radioactive material the radioactivity falls with time; the activity is proportional to the amount of substance present.

Decay follows an exponential law. At any time t the number (N) of original nuclei present is given by:
$$N = N_0 \exp -\gamma t$$
where N_0 is the original number and γ is a constant (the *decay constant* or *disintegration constant*) for a given nuclide. γ is related to the half-life by:
$$T = 0.693\ 15\ \gamma$$

decay constant *See* decay.

deci- Symbol: d A prefix denoting 10^{-1}. For example, 1 decimetre (dm) = 10^{-1} metre (m).

decibel Symbol: dB A unit of power level, usually of a sound wave or electrical signal, measured on a logarithmic scale. The threshold of hearing is taken as 0 dB in sound measurement. Ten times this power level is 10 dB. The fundamental unit is the *bel*, but the decibel is almost exclusively used (1 db = 0.1 bel).

A power P has a power level in decibels given by:
$$10 \log_{10}(P/P_0)$$
where P_0 is the reference power.

declination Symbol: ϵ The angle between the geographic and magnetic meridians at a given point on the Earth's surface. Allowance must be made for this when using a compass to find geographical direction. At any point on the Earth's surface the declination changes slowly with time.
See also Earth's magnetism, magnetic variation.

defect An irregularity in the ordered arrangement of particles in a crystal lattice. There are two main types of defect in crystals. *Point defects* occur at single lattice points. A *vacancy* is a missing atom; i.e. a vacant lattice point. Vacancies are sometimes called *Schottky*

defects. An *interstitial* is an atom that is in a position that is not a normal lattice point. If an atom moves off its lattice point to an interstitial position the result (vacancy plus interstitial) is a *Frenkel defect*. All solids above absolute zero have a number of point defects, the concentration of defects depending on the temperature. Point defects can also be produced by strain or by irradiation. *Dislocations* (or *line defects*) are also produced by strain in solids. They are irregularities extending over a number of lattice points along a line of atoms.

definition The sharpness of focus of an image. No image can be perfectly sharp because of diffraction, even with a perfect optical system. *See also* resolving power.

deflection magnetometer *See* magnetometer.

degaussing The neutralization of an object's magnetic field by the use of an equal and opposite field. It is used in colour television sets to neutralize the Earth's magnetic field — current-carrying coils prevent the formation of colour fringes on the image. A similar system is used to neutralize the magnetization of ships to prevent them from detonating magnetic mines.

degenerate Describing different quantum states that have the same energy. For instance, the three p orbitals in an atom all have the same energy but different values of the magnetic quantum number m. Differences in energy occur if a magnetic field is applied — the degeneracy is then said to be 'lifted'.

degree 1. An interval on a scale of measurement, such as a temperature scale.
2. A unit of angle; $1/360$ of a complete turn.

degrees of freedom The independent ways in which particles can take up energy. In a monatomic gas, such as helium or argon, the atoms have three translational degrees of freedom (corresponding to motion in three mutually perpendicular directions). The mean energy per atom for each degree of freedom is $kT/2$, where k is the Boltzmann constant; the mean energy per atom is thus $3kT/2$.

A diatomic gas also has two rotational degrees of freedom (about two axes perpendicular to the bond) and one vibrational degree. The rotations also each contribute $kT/2$ to the average energy. The vibration contributes kT ($kT/2$ for kinetic energy and $kT/2$ for potential energy). Thus, the average energy per molecule for a diatomic molecule is $3kT/2$ (translation) + $2kT/2$ (rotation) + kT (vibration) = $7kT/2$.

Linear triatomic molecules also have two significant rotational degrees of freedom; non-linear molecules have three. For non-linear polyatomic molecules, the number of vibrational degrees of freedom is $3N - 6$, where N is the number of atoms in the molecule.

The molar energy of a gas is the average energy per molecule multiplied by the Avogadro constant. For a monatomic gas it is $3RT/2$, etc. See also specific thermal capacity.

demagnetization The removal of the magnetic field of a magnet. In demagnetization, the domains are disordered so that they lie in a random pattern. This can occur as a result of rough treatment, or by raising the temperature of the specimen and allowing it to cool while lying in an East–West direction. The most efficient method is to place the specimen in a coil through which is passed an alternating current. The current is then reduced to zero, removing the magnetism. A similar effect is obtained by taking the specimen out of a coil carrying an alternating current.

demodulation (detection) The process in which the original modulating signal is reproduced in its original form from a modulated carrier wave by a circuit. See also detector, modulation.

densitometer An instrument for measuring the density of exposure of a photographic emulsion. Densitometers have a light source and photocell to measure the transmission (or reflection) of the sample. They are used in obtaining quantitative measurements from photographic records of spectra, diffraction experiments, etc.

density 1. Symbol: ρ The mass of a substance per unit volume. The units are kg m^{-3}.
2. See relative density.
3. See charge density.

depolarizer A substance used in a voltaic cell to prevent polarization. Hydrogen bubbles forming on the electrode can be removed by an oxidizing agent, such as manganese dioxide. See also polarization (electrolytic).

depth of field (depth of focus) The object distance range of an optical system (e.g. a camera) that gives acceptably sharp images. The smaller the aperture, the greater the depth of field.

depth of focus See depth of field.

derived unit A unit defined in terms of base units, and not directly from a standard value of the quantity it measures. For example, the newton is a unit of force defined as a kilogram metre second^{-2} (kg m s^{-2}). See also SI units.

destructive interference See interference.

detector 1. (demodulator) An electronic circuit for extracting the original modulating signal from a carrier wave. See also modulation.
2. A device that responds to any physical effect, used to indicate the presence of a signal or to measure it. An example is an aerial inducing a voltage at the frequency of a received radio wave.

deuterated See deuterium.

deuterium Symbol: D A naturally occurring stable isotope of hydrogen in which the nucleus contains one proton and one neutron. The atomic mass is thus approximately twice that of ^{1}H; deuterium is known as 'heavy hydrogen'. Chemically it behaves almost identically to hydrogen, forming analogous compounds, although reactions of deuterium compounds are often slower than those of the corresponding ^{1}H compounds. Its physical properties differ slightly from the properties of ^{1}H. Deuterium oxide, D_2O, is called 'heavy water'; it is used as a moderator and coolant in some types of nuclear reactor.

Chemical compounds in which deuterium atoms replace the usual ^{1}H atoms are said to be *deuterated*.

The natural abundance of deuterium is somewhat below 0.015% but this varies slightly with source.

deuteron A nucleus of a deuterium atom; i.e. a combination of a proton and a neutron.

deviation The turning of a ray during reflection or refraction. The *angle of deviation* is the angle through which the ray's direction changes — an undeviated ray has zero angle of deviation. In the case of a reflecting surface, the angle of deviation (d) can vary between almost zero and 180°. A refracting surface can produce values of d between zero and the complement of the critical angle (180° - c), where c is the critical angle.

The angle of deviation produced by a prism varies with the angle of incidence and has a minimum value (D) when the ray passes symmetrically through the prism. D is related to the prism's refractive constant (n) and its angle (A):

$$n = \sin[(A + D)/2]/\sin(A/2)$$

This equation can be used for finding the refractive constant of the material of the prism using a spectrometer to measure D.

dew Water droplets that condense from air onto a cooler surface, such as may happen outdoors at night. After dark, a leaf, for example, will radiate and become cooler. If it is surrounded by calm warm moist air, condensation will occur, just as it does on a cold window pane in a relatively warm humid room. If the surface temperature is below the freezing temperature of water, the condensation will be as frost. *See also* relative humidity.

Dewar flask (vacuum flask) A double-walled container of thin glass with the space between the walls evacuated and sealed to stop conduction and convection of energy through it. The glass is often silvered to reduce radiation.

dew temperature (point) *See* relative humidity.

dextrorotatory *See* optical activity.

diamagnetism The magnetic behaviour of substances (e.g. bismuth and lead) with a negative susceptibility. The relative permeability is less than one. Usually the value of susceptibility is small for diamagnetic substances and relative permeability is slightly less than one. Diamagnetism results from the motion of electrons in atoms; under the influence of an applied field these change so as to oppose the field. The magnetization of a diamagnetic material is thus opposite to that of a paramagnetic or ferromagnetic material. If a bar of a diamagnetic substance is placed in a non-uniform field it settles with its axis perpendicular to the field.

Diamagnetism is a weaker effect than paramagnetism or ferromagnetism. It is shown by all substances and is independent of temperature, although in some materials it is masked by other types of magnetic behaviour. *See also* magnetism.

diaphragm A device used in optical instruments to reduce or control the aperture of the system. An *iris diaphragm* is a variable aperture made up of a number of movable overlapping plates. Diaphragms (or *stops*) are used to control the light admitted. They are also employed to reduce aberration in lens and mirror systems by restricting the light to the region close to the axis of the system. *See also* aberration.

diascope An optical projector for translucent objects (e.g. slides). The name is rather old-fashioned.

diatomic Describing a molecule that consists of two atoms. Hydrogen (H_2), oxygen (O_2), and nitrogen (N_2) are examples of diatomic gases.

dichroic medium A birefringent material in which radiation polarized in one plane is freely transmitted, but radiation polarized perpendicular to this is absorbed. Tourmaline is a natural mineral with this property; Polaroid is a synthetic dichroic substance.

dielectric A nonconductor of electricity. The name is usually used where electric fields are possible, as with the insulating material between the plates of a capacitor. *See also* relative permittivity.

dielectric constant An early term for relative permittivity.

dielectric heating The heating effect caused by the rapid alternation of the electric field across an insulator (dielectric) (as with the material between the plates of a capacitor in an a.c. circuit). The effect is used as a heating technique for certain materials. The loss of useful energy that results is *dielectric loss*.

dielectric loss *See* dielectric heating.

dielectric strength The maximum electric field that a dielectric can withstand without causing it to break down (and pass a spark discharge). It is usually measured in volts per millimetre.

diffraction The effects of an obstacle on passing radiation, making the shadow edge less sharp than expected. Some radiation bends round the obstacle into the shadow region; in some parts of the non-shadow region the radiation is reduced in intensity. Diffraction is observed with all waves, particularly if the obstacle is comparable in size to the wavelength. Under these conditions, fringes (diffraction patterns) are formed. *See also* Fraunhofer diffraction, Fresnel diffraction.

diffraction grating The normal form is a sheet of glass with fine parallel close grooves (lines) scraped across it. Diffraction around the edges of each line produces patterns of light and dark 'fringes' at different angles. As the fringe spacing depends on wavelength, diffraction gratings are used to make spectra of incident light. Maximum intensity is given at values of θ such that:
$$\sin\theta = m\lambda/d \, (m = 0, 1, 2, \text{etc.})$$
The value of m is the *order* of the spectral line concerned; d is the distance between neighbouring grating lines. As $\sin\theta$ cannot exceed 1, the value of λ/d determines the number of orders of spectra obtainable with the grating.

diffuse reflection *See* reflection.

diffuse refraction *See* refraction.

diffusion 1. Movement of a gas, liquid, or solid as a result of the random thermal motion of its particles (atoms or molecules). A drop of ink in water, for example, will slowly spread throughout the liquid. Diffusion in solids occurs very slowly at normal temperatures. *See also* Graham's law.
2. The random spreading out of a beam of radiation on reflection or transmission. Diffusion occurs when a beam of light passes through a translucent medium (e.g. wax or frosted glass) or is reflected by a rough surface (e.g. paper or matt paint). *See also* reflection, refraction.

diffusion cloud chamber *See* cloud chamber.

dihedral Formed by or having two intersecting planes or sides. For example, the angle between two intersecting plane faces of a crystal is a *dihedral angle*.

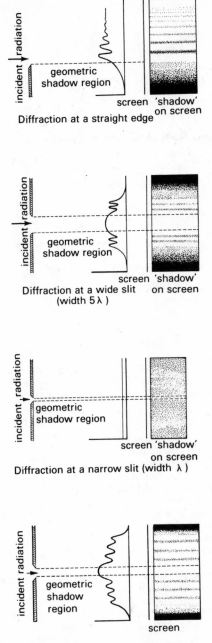

Diffraction at a straight edge

Diffraction at a wide slit
(width 5 λ)

Diffraction at a narrow slit (width λ)

Diffraction at a slit (width 2½ λ)

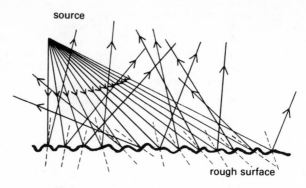

source

rough surface

Diffusion of radiation

Dimensions and units of some physical quantities

Quantity	Dimension	Unit	Normal unit
mass	[M]	kg	kg
length	[L]	m	m
time	[T]	s	s
area	[L²]	m²	m²
volume	[L³]	m³	m³
density	[ML⁻³]	kg m⁻³	kg m⁻³
acceleration	[LT⁻²]	m s⁻²	m s⁻²
force	[MLT⁻²]	kg m s⁻²	N
pressure	[ML⁻¹T⁻²]	kg m s⁻²m⁻²	Pa
momentum	[MLT⁻¹]	kg m s⁻¹	(N s)
pulsatance	[T⁻¹]	s⁻¹	Hz

dimensional analysis A technique for checking the validity of a solution to a problem, a unit, or an equation in physics. It depends on the fact that, once a unit system has been defined fully (as in SI), each quantity can be expressed in terms of the base quantities in only one way. Here those base quantities are called *dimensions*. The table shows, first, the three dimensions used in mechanics, and, second, some other quantities in terms of these. For instance, the relationship between pressure, force, and area can be checked. Force divided by area has dimensions MLT⁻²/L², i.e. dimensions ML⁻¹T⁻², as for pressure.

dimensions *See* dimensional analysis.

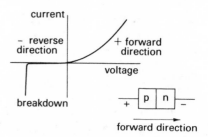

current

– reverse direction

+ forward direction

voltage

breakdown

forward direction

diode An electronic device with two electrodes that exhibits rectifying action when a potential difference is applied. Current flows for one direction of the potential (the forward direction), but when the potential is reversed the current is normally zero or very small. There are two principal types of diode:

1. The *semiconductor diode* consists of a single p-n junction. The current across the junction increases exponentially with voltage in the forward direction; in the reverse direction there is only a very small (leakage) current until breakdown occurs.

2. The *thermionic diode* consists of a heated negative cathode emitting electrons (by thermionic emission) in a vacuum with a positive anode, which accepts electrons when a positive potential is applied to it. The emitted electrons form a 'cloud' around the cathode, called the *space charge*. Since electrons repel each other owing to their like charges, the space charge limits the current as the voltage is increased. When the voltage is high enough the current reaches a maximum saturation value.

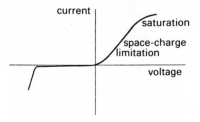

Current-voltage characteristic

Some other types of semiconductor diode are the light-emitting diode (LED), the tunnel diode, and the Zener diode.
See also rectifier, semiconductor, thermionic valve.

dioptre (diopter) Symbol: D The unit of power of a lens or mirror, equal to the reciprocal of the focal distance in metres. A 5 D lens has a focal distance of 0.2 m; a diverging mirror of focal distance 1250 mm has a power of -0.8 D.
Power measures the ability of a reflector, refracting surface, or lens to converge a parallel beam. The dioptre is often now called the radian per metre (rad m^{-1}).

dip, angle of *See* inclination.

dip circle *See* inclinometer.

direct current (d.c.) Electric current in one direction only. *Compare* alternating current.

disappearing-filament pyrometer *See* optical pyrometer.

discharge, electrical The passage of an electric current through a gas at low pressure or a dielectric. It is usually accompanied by luminous effects. The gas molecules become ionized and the recombination of oppositely charged ions gives rise to the luminous effects.
See also canal rays, corona discharge, electric arc, electric spark, glow discharge, positive column.

discriminator A circuit that selects certain electrical signals and rejects others, according to a particular property of the signal, such as amplitude, frequency, or phase. For example, a pulse height analyser selects signals having a particular amplitude range, and a frequency discriminator picks out different frequencies.

disintegration The breaking up of unstable nuclei into two or more fragments, either spontaneously or as a result of bombardment by fast moving particles. The disintegration of ^{7}Li by proton bombardment in Cockcroft and Walton's experiment is an example of the latter type; two ^{4}He nuclei (alpha particles) are produced.

disintegration constant *See* decay.

dislocation *See* defect.

dispersion During refraction, rays can be deviated (changed in direction) by angles related to the refractive constant of the medium. However, the refractive constant of any medium varies with the wavelength of the transmitted radiation. Thus, if a narrow beam of radiation of mixed wavelengths is deviated by refraction, the angle of deviation will vary with wavelength. This is called dispersion; the radiation is spread out. It can be used to produce the colour spectrum of white light using a prism. It can also be a problem, causing chromatic aberration in lenses.

Dispersion is actually an effect in which radiations having different wavelengths travel at different speeds in the medium. Normally the variation of speed (and thus of refractive constant) with wavelength is continuous. At absorbed wavelengths, however, the curve is discontinuous, leading to *anomalous dispersion*.

dispersion forces *See* van der Waals forces.

dispersive power Symbol: ω A measure of the variation for a medium of refractive constant n with wavelength λ. It is given by the ratio:
$$\omega = (n_b - n_r)/(n_y - 1)$$
Here, n_b, n_r, and n_y are absolute refractive constants for blue, red, and yellow light. More accurate values are obtained by using three monochromatic spectral lines of known wavelength: F (blue), C (red), and D (yellow).

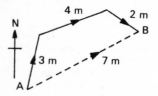

distance from A to B is 9 m
displacement is 7 m at
a bearing of N 73°E (073°)

Displacement

displacement Symbol: s The vector form of distance, measured in metres (m) and involving direction as well as magnitude.

dissipation The removal of energy from

a system to overcome some form of resistive force (mechanical or electrical). Without resistance (as in motion in a vacuum and the current in a superconductor) there can be no dissipation. Dissipated energy normally appears as thermal energy.

distance Symbol: d The length of the path between two points. The SI unit is the metre (m). Distance may or may not be measured in a straight line. It is a scalar; the vector form is displacement.

distance ratio (velocity ratio) For a machine, the ratio of the distance moved by the effort in a given time to the distance moved by the load in the same time. *See also* machine.

distortion *See* aberration.

divergent Moving apart. *See* convergent.

diverging lens A lens that can refract a parallel beam into a divergent beam. Diverging lenses used in air are thinner in the middle than at the edge. They may be biconcave, plano-concave, or concavo-convex in shape. As these lenses have negative power, they are sometimes called *negative lenses*.
Compare converging lens. *See also* lens.

diverging mirror (diverging reflector) A mirror that can reflect a parallel beam into a divergent beam. Diverging reflectors always have convex surfaces. As these reflectors have negative power, they are sometimes called *negative mirrors*.
Compare converging mirror. *See also* mirror.

diverging reflector *See* diverging mirror.

domain A region inside a ferromagnetic material in which all the atomic magnetic fields point the same way. When the material is unmagnetized, the directions of the separate domains are random; when a strong external magnetic field is applied, the domains line up in the same direction. The domains with magnetic axes nearly parallel to that of the magnetizing field grow at the expense of those perpendicular to the field. The substance now has a resultant magnetic moment. The domains can grow until the magnetic axes are in line with the external field if this is strong enough. The material is then magnetically saturated. The presence of magnetic domains can be shown by Bitter patterns, the Barkhausen effect, and by various other techniques.

donor *See* semiconductor.

dopant *See* semiconductor.

doping *See* semiconductor.

Doppler broadening The increase in the width of a spectral line as a result of the motion of the particles emitting or absorbing the radiation. *See* monochromatic radiation.

Doppler effect 1. An apparent change in the frequency of a wave motion, caused by relative motion between the source and the observer. It is found with sound waves: as the source and observer move together the apparent frequency of the sound is higher than that produced; as they move apart it is lower. An example is that of an approaching motorbike — the pitch of the engine sound appears to drop suddenly as it passes.

The apparent frequency (f) of sound as a result of relative motion between source and observer in a direct line is given by:
$$f = (V + U_O)f/(V - U_S)$$
where V is the speed of sound, U_O the observer's velocity, U_S the source velocity, and f the source frequency.

2. A similar effect for electromagnetic radiation. The radiation emitted by a source with a motion component towards the observer will have increased frequency: if the source is moving away from the observer, the observed frequency is reduced. This has great significance in astronomy. Thus the Doppler effect on radar pulses reflected by the planets gives information on their rota-

tion. The displacement of lines in the spectra of stars and galaxies (i.e. the red shift or the blue shift) can tell the astronomer how these bodies are moving.

For electromagnetic radiation the theory of the effect is not the same as for sound. This is because there is no fixed medium to give a frame of reference. Relativity theory gives the following result:
$$v = v_0 \sqrt{[(1 - v/c)/(1 + v/c)]}$$
Here v is the observed frequency, v_0 is the emitted frequency; v is the speed of separation of source and observer, related to the speed of light c. If v is small compared to c the expression simplifies to:
$$v = v_0(1 - v/c)$$

Doppler shift *See* red shift.

dose The quantity of energy from ionizing radiation (e.g. X-rays, alpha, beta, or gamma radiation) absorbed by a living organism. Doses are measured in roentgens, rems, or rads. Doses of ionizing radiation result from natural background radioactivity, X-ray diagnosis, occupational exposure, and release of radioactive material into the environment (e.g. from nuclear explosions). Maximum permissible doses for various organs, for people working with radioactive materials and for the general public, are recommended by the International Commission on Radiological Protection. Controlled doses of ionizing radiation may also be used medically for diagnosis and treatment.

dot product *See* scalar product.

double refraction *See* birefringent crystal.

double slit experiment *See* Young's double slit.

doublet A compound lens with two elements, often in contact. The purpose is usually to reduce aberrations. If the lenses are in contact, the power of the doublet is the sum of the powers of the elements.

Thus, $P = P_1 + P_2$

or, $1/f = 1/f_1 + 1/f_2$ (f is focal distance).

A 5 D lens combined with a -2.5 D lens forms a doublet of power 2.5 D, or focal distance 400 mm.

drain *See* field effect transistor.

dry cell A voltaic cell in which the electrolyte is in the form of a jelly or paste. Dry cells are used for flashlights and other portable appliances. *See also* Leclanché cell.

duality Wave-particle duality; the property of behaving like waves in some situations and like particles in others. Electromagnetic radiation behaves as a wave motion in diffraction, for instance, but as particles (photons) in the photoelectric effect. *See also* de Broglie wave.

Dulong and Petit's law The molar thermal capacity of a solid element is approximately equal to $3R$, where R is the gas constant (25 J k^{-1} mol^{-1}). The law applies only to elements with simple crystal structures at normal temperatures. At lower temperatures the molar heat capacity falls with decreasing temperature (it is proportional to T^3). Molar thermal capacity was formerly called *atomic heat* — the product of the atomic weight and specific thermal capacity.

dynamic equilibrium *See* equilibrium.

dynamics The study of how objects move under the action of forces. This includes the relation between force and motion and the description of motion itself. *See* equations of motion, Newton's laws of motion.

dynamo A small device for converting mechanical energy into electrical energy. The motion of a coil rotating in a magnetic field induces an alternating current in the windings. This current itself causes a turning force opposing the initial motion. To keep the dynamo turning, an energy input is required from an external source, such as a petrol or diesel engine. Many bicycles have a small dynamo built into the hub of the wheel (or driven by the tyre) to power the cycle lamps. *See also* generator.

dyne Symbol: dyn The former unit of force used in the c.g.s. system. It is equal to 10^{-5} N.

E

Earth inductor A type of fluxmeter for measuring large uniform fields (normally that of the Earth). A large coil of known area and number of turns is flipped through 180° and the induced e.m.f. is measured using a ballistic galvanometer.

earthing (grounding) The connection of an electrical conductor to a large conducting body, such as the Earth, which is used as the zero in the scale of electrical potential. Earthing a conductor causes charge to flow until it too is at zero potential. In electrical appliances, any metal casings are usually earthed for safety, so that if the casing accidentally becomes connected to the live supply it will not give a shock to anyone who touches it. Often part of a circuit is deliberately earthed to fix its potential.

Earth's magnetism A magnetic field can be observed at any point on or near the Earth's surface. This field is rather like that which would be obtained if a giant bar magnet were present inside the planet; it is closer to the field of a magnetized steel sphere. A magnetic compass can be used to show the direction of the field — of great importance in navigation. The Earth's north magnetic pole is currently off the northern coast of Canada; the south magnetic pole is on the edge of Antarctica. The strength and direction of the Earth's field at any point are defined in terms of three magnetic elements — the inclination, the declination, and the horizontal component of the field strength. These elements are not constant.

The source of the Earth's field is not known. A feasible theory is that it is due to electric currents circulating in the molten nickel–iron core.
See magnetic variation.

echelon A type of diffraction grating consisting of a number of equal thin glass sheets stacked on a slant. In use the light is reflected from the stepped side of the stack; *d* in the grating equation is very large, so that very high spectral orders are possible.

echo The result of reflection of sound waves (or other waves), usually by a hard surface. This causes an observer to detect two signals; the first a direct signal and the second a fainter reflected signal. The delay between the two indicates the distance of the reflector.

echo sounder See sonar.

eclipse The 'hiding' of one heavenly body behind another. The eclipsed object, the eclipsing object, and the observer are in a straight line.

In a *solar eclipse*, the Moon appears to pass across the Sun, so that the Moon's shadow sweeps over the Earth's surface. If the Moon appears to cover the Sun completely, there is a *total eclipse* of the Sun — the observer is in the umbra of the Moon's shadow. If the Moon only partly covers the Sun, there is a *partial eclipse* — the observer is in the penumbra of the shadow. Solar eclipses can happen only at New Moon; the Moon's shadow is narrow, so total eclipses of the Sun are rare at any one place on the Earth. It is a coincidence that Sun and Moon appear to be of similar size from the Earth. However as neither the Earth-Moon distance nor the Sun-Earth distance is constant, the sizes of Sun and

Moon seen from the Earth do vary. Sometimes there is an *annular eclipse*, when the Moon is not 'big' enough to cover the Sun's disc. Then the vertex of the lunar umbra does not reach the Earth's surface.

As the Earth's shadow is larger than the Moon's, *lunar eclipses* are more commonly seen than solar eclipses. Lunar eclipses can occur only at Full Moon, when the Earth's shadow sweeps the Moon's surface. They also last longer than solar eclipses, as Earth and Moon are now moving in the same direction through space. Another big difference is that lunar eclipses do not clearly show umbra and penumbra. This is because sunlight is refracted into the umbra cone by the Earth's atmosphere, acting as a ring-shaped lens. This refraction also explains the reddish colour of the Moon during eclipse.

eddy-current heating See induction heating.

eddy currents Induced currents set up in a conductor by a changing magnetic field. They occur in transformers and other electrical devices. The currents produce a heating effect corresponding to a loss of useful energy (*eddy-current loss*). Metal cores in electrical machines are usually laminated (built of thin sheets) to reduce such losses; the surface layers between the laminations have high electrical resistance.

One useful application of eddy currents is in damping moving-coil instruments — the coil is wound on a piece of soft iron. When the coil moves in the instrument's magnetic field, eddy currents are induced in the iron so as to oppose the motion, causing the coil to settle quickly. As Lenz's law would predict, eddy currents always flow so as to oppose the effect producing them.

Edison cell See nickel–iron accumulator.

effective resistance The resistance of part of an electric circuit in which there is an alternating current. It is the power dissipated divided by the square of the current. If power is in watts and current in amperes, the units are then ohms. Effective resistance expresses a combination of two effects: the resistance from scattering of electrons by lattice atoms (as for d.c.) and the resistance from the magnetic effects of the changing current.

effective value See root-mean-square value.

efficiency Symbol: η A measure used for processes of energy transfer; the ratio of the useful energy produced by a system or device to the energy input. Similarly, the efficiency is equal to the useful power output divided by the power input. There is no unit; however efficiency is often quoted as a percentage. There are several branches of physics in which efficiency is used:
1. For machines, the efficiency is the force ratio divided by the distance ratio. See machine.
2. For an electric cell delivering current through a load resitance (R), the efficiency is given by $\eta = R(R + r)$, where r is the internal resistance of the cell. Note that the higher the value of R, the higher the efficiency.
3. For an electric motor, the efficiency is the mechanical power developed divided by the electrical power input.
4. For a reversible heat engine, the efficiency is given by $\eta = (T_1 - T_2)/T_1$, T_1 being the source temperature and T_2 the 'sink' temperature.

effusion The flow of gas molecules through relatively large holes.

EHF See extremely high frequency.

Einstein's equation See photoelectric effect.

Einstein shift See red shift.

elastance The reciprocal of capacitance. It is measured in farad^{-1} (F^{-1}).

elastic collision A collision for which

the restitution coefficient is equal to one. Kinetic energy is conserved during an elastic collision. See also restitution, coefficient of.

elasticity The tendency of a material to return to its original dimensions after a deforming stress has been removed. Up to a certain point (the proportional limit) the material obeys Hooke's law. The strain produced is proportional to the stress and the sample returns to its original dimensions if the stress is removed. Above the proportional limit the material no longer obeys Hooke's law, i.e. stress is not directly proportional to strain. At slightly higher stresses the *elastic limit* is reached. Above this the sample no longer returns to its original dimensions after the stress is removed; i.e. a permanent deformation occurs. The *yield point* is the point at which the material begins to 'flow' — i.e. the strain increases with time (up to the breaking point) without further increase in the stress.

elastic limit See elasticity.

elastic modulus The ratio of the stress on a body to the strain produced. There are various moduli of elasticity depending on the type of stress applied. The *Young modulus* (symbol: E) refers to tensile or compressive stress. It is the force per unit cross sectional area divided by the fractional elongation of the sample. The bulk modulus (symbol: K) is used for a bulk stress — i.e. an overall pressure on the body. It is the compressive (or tensile) force per unit area divided by the fractional change in volume. The *shear modulus* (or *rigidity modulus*) (symbol: G) refers to a shearing (or twisting) stress. It is the tangential force per unit area divided by the angular deformation. Elastic moduli have the same units as stress (newtons per square metre).
See also elasticity.

electret A piece of permanently electrified material, having a positive charge at one end and a negative charge at the other.

electrical energy Energy resulting from the position of an electric charge in an electric field.

electric arc A luminous electrical discharge with a high current density, occurring between two electrodes. In certain types of electric arc the heating of the electrodes vaporizes the electrode material, leading to an emission spectrum of this substance in the light. Electric arcs are thus used as sources for spectrographic analysis. The heating effect of an electric arc is also used in arc welding. Arc lamps give a high intensity light source produced by an arc between carbon electrodes.

electric bell A bell in which the to-and-fro movement of the hammer is produced by an electromagnet. In a direct-current (battery-operated) bell, the electromagnet consists of two coils wound on two cores connected by an iron yoke. When the button is pressed to close the circuit, the coils are energized and attract an iron bar (armature). The armature movement breaks the circuit, the coil de-energizes again, and the armature moves back (by a spring) and closes ('makes') the circuit again, and so on, as long as the switch is closed. An alternating-current bell does not have this type of *make-and-break contact*. The armature is a permanent magnet, which is alternately attracted and repelled by the reversing poles of an electromagnet through which the alternating current passes.

electric charge *See* charge.

electric constant *See* permittivity.

electric current *See* current.

electric displacement Symbol: D The electric flux density in a dielectric material:
$$D = \epsilon_r\epsilon_0 E$$
where E is the electric field strength in

free space and $\epsilon_r\epsilon_0$ the permittivity of the material.

electric field A region in which a force would be exerted on an electric charge. Like magnetic fields, electric fields can be represented by lines of force. The electric field is completely defined in magnitude and direction at any point by the force upon a unit positive charge situated at that point. Electric fields can be produced by electric charges or by changing magnetic fields.

electric field strength (electric intensity) Symbol: E A measure of the strength of an electric field at a point, defined as the force per unit charge on an electric charge placed at that point: i.e. $E = F/Q$. The electric field strength is the gradient of the electrical potential at a point ($-dV/dr$); the unit used is the volt per metre ($V\,m^{-1}$), equal to the newton per coulomb ($N\,C^{-1}$).

electric flux Symbol: Ψ A measure of an electric field in a vacuum or in a dielectric. The flux in a field can be represented by the number of lines of electric force through a given perpendicular area. The *electric flux density* at a point is the number per unit area at that point. Electric flux density is related to electric displacement.

electric image A method of solving electrostatic problems. A charge $+Q$ a certain distance from a conducting surface acts as if there were an 'image' of the charge of magnitude $-Q$ positioned on the opposite side of the surface. For a plane conductor the image is as far behind the surface as the charge is in front.

electric intensity *See* electric field strength.

electricity The nature and effects of moving or stationary electric charges.

electric motor A device for converting electrical energy into motion. Electric current passed through a coil forms an electromagnet, which causes the coil (or *rotor*) to rotate in a second magnetic field. The second field may be produced by a permanent magnet or another electromagnet. The rotating motion itself generates an electromotive force (back e.m.f.) in the coil, in the opposite direction to the current flow. A certain amount of electrical energy therefore has to be used to maintain the current in the coil. When the motor is also turning an external mechanical load, greater forces have to be overcome and therefore a greater coil current is needed to turn the rotor at the same speed. An increase in mechanical power output therefore draws more electrical power from the supply. Electric motors can be designed to work with direct-current supply or, more commonly, with alternating current. *See also* induction motor, synchronous motor.

electric polarization *See* polarization.

electric potential Symbol: V The potential at a point in an electric field is the energy required to bring unit electric charge from infinity to the point. The unit of electric potential is the volt, V. *See also* potential difference.

electric power *See* power.

electric spark A discharge of electricity in a gas accompanied by light and sound. The discharge occurs between two points of opposite high electric potentials.

The spark can only occur when the potential difference exceeds a certain value, called the sparking potential. The sudden expansion and then contraction of the gas after it has been heated by the spark gives rise to the characteristic clicking sound.

electrochemical equivalent Symbol: z The mass of an element released from a solution of its ion when a current of one ampere flows for one second during electrolysis.

electrochemical series *See* electromotive series.

electrochemistry The study of the formation and behaviour of ions in solutions. It includes electrolysis and the generation of electricity by chemical reactions in cells.

electrode Any part of an electrical device or system that emits or collects electrons or other charge carriers. An electrode may also be used to deflect charged particles by the action of the electrostatic field that it produces.

electrodeposition The process of depositing a layer of solid (metal) on an electrode by electrolysis. Positive ions in solution gain electrons at the cathode and are deposited as atoms. Copper, for instance, can be deposited on a metal cathode from copper sulphate solution.

electrode potential A measure of the tendency of a metal to lose electrons to a surrounding solution. It is the potential difference, at equilibrium, between the metal and a solution of its ions.

The electrode potential cannot be measured directly and is usual!y compared against that of a hydrogen electrode. Metals that have a greater tendency than hydrogen to form positive ions have a positive electrode potential. Electrode potentials are quoted as measured under standard conditions. *See also* half cell.

electrodynamics The study of the relationship between mechanical forces and magnetic and electric forces. Such effects are the basis of, for example, the electrical generator and the electric motor.

electroluminescence *See* luminescence.

electrolysis The production of chemical change by passing charge through certain conducting liquids (electrolytes). The current is conducted by migration of ions — positive ones (cations) to the cathode (negative electrode), and negative ones (anions) to the anode (positive electrode). Reactions take place at the electrodes by transfer of electrons to or from them.

In the electrolysis of water (containing a small amount of acid to make it conduct adequately) hydrogen gas is given off at the cathode and oxygen is evolved at the anode. At the cathode the reaction is:

$$H^+ + e^- \rightarrow H$$
$$2H \rightarrow H_2$$

At the anode:

$$OH^- \rightarrow e^- + OH$$
$$2OH \rightarrow H_2O + O$$
$$2O \rightarrow O_2$$

In certain cases the electrode material may dissolve. For instance in the electrolysis of copper(II) sulphate solution with copper electrodes, copper atoms of the anode dissolve as copper ions

$$Cu \rightarrow 2e^- + Cu^{2+}$$

electrolyte A liquid containing positive and negative ions, that conducts electricity by the flow of those charges.

Electrolytes can be solutions of acids or metal salts ('ionic compounds'), usually in water. Alternatively they may be molten ionic compounds — again the ions can move freely through the substance. Liquid metals (in which conduction is by free electrons) are not classified as electrolytes. See also electrolysis.

electrolytic capacitor A type of capacitor consisting of two metal plates, one with a very thin oxide layer, immersed in a suitable electrolyte. Electrolytic capacitors provide a high capacitance for a small size. It is important that they are always charged up with the correct plates positive and negative, otherwise the oxide film may break down.

electrolytic cell See cell, electrolysis.

electrolytic separation The separation of isotopes by electrolysis. See isotope separation.

electromagnet A coil of wire wrapped around a core of soft iron. When there is a current in the wire, a magnetic field

results; the core becomes magnetized. The core loses its magnetism when the current is switched off. Electromagnets are used in telephones, electric bells, and many other devices. See also solenoid.

electromagnetic induction The induction of an electromotive force (e.m.f.) between the ends of a conductor by change in relation to an external magnetic field. The 'laws of electromagnetic induction' follow:

(1) (Faraday) An electromotive force is induced in a conductor when there is a change in the magnetic field around it.

(2) (Faraday) The electromotive force induced is proportional to the rate of change of the field.

(3) (Lenz) The direction of the induced electromotive force tends to oppose the change.

Electromagnetic induction is used in generators, transformers, microphones, and many other devices. The effect can also cause problems, as in the formation of eddy currents.

The magnitude of the induced e.m.f. is proportional to the rate of change of total flux linkage, Φ. This is known as Neumann's law. Thus,

$$E = -d\Phi/dt$$

where E is the e.m.f. in volts, Φ the flux linkage in webers, and t the time in seconds.

See also mutual induction, self-induction.

electromagnetic interaction See interaction.

electromagnetic moment See magnetic moment.

electromagnetic pump. A pump with no moving parts, used for circulating electrically conducting fluids. A direct current is passed between two electrodes inserted in the fluid circuit and a constant magnetic field is directed perpendicular to this current. There is a force on the fluid mutually perpendicular to these two directions, and therefore parallel to the direction of flow,

electromagnetic radiation

thus propelling the fluid. One of its applications is in pumping liquid sodium coolant through nuclear reactors.

electromagnetic radiation Energy propagated by vibrating electric and magnetic fields. Electromagnetic radiation forms a whole electromagnetic spectrum, depending on frequency and ranging from high-frequency radio waves to low-frequency gamma rays. Electromagnetic radiation can be considered as waves (*electromagnetic waves*) or as streams of photons. The frequency and wavelength are related by $\lambda\nu = c$, where c is the speed of light. The energy carried depends on the frequency. *See also* photoelectric effect.

electromagnetic spectrum As electromagnetic radiation has wave properties, it must be possible to have a whole series that varies in frequency. This is in fact the case. Visible light is only a small part of that spectrum. The other radiations behave in the same way as light — showing reflection, refraction, absorption, diffraction, interference, and polarization in the right circumstances. All travel at the same speed through a vacuum — close to 3×10^8 m s^{-1} (300 000 km s^{-1}).

The table outlines the spectrum, with indication of the frequencies and wavelengths of each sector. Note that this *is* a spectrum — there are no sharp boundaries between the types, but a gradual change in behaviour as the frequency changes.

electromagnetic units *See* c.g.s. system.

electromagnetic waves *See* electromagnetic radiation.

electrometer An electronic instrument for measuring low potential differences and currents, while drawing very little current from the circuit. Electrometers are d.c. amplifiers with very high input impedances. Early types (*valve voltmeters*) used special valve circuits for amplifying steady signals. In *vibrating-reed electrometers* a capacitor is used with one plate mechanically vibrated at constant frequency. A steady low potential difference can thus be converted into an alternating one, which can then be amplified. More modern types of electrometer use transistor circuits. Electrometers are often used for measuring very low currents (less than 10^{-9} A) by passing the current through a standard high resistance. Ionization currents are measured in this way.

electromotive force (e.m.f.) Symbol: E A measure of the energy supplied by a source of electric current. It is the energy involved in taking unit charge round the circuit in the opposite direction to the electromotive force. It can also be defined as the power supplied by a source of electric current per unit current. The unit of e.m.f. is the volt. There are a number of ways in which an e.m.f. can be generated — for example, by an electric cell, by a coil moving in a

Electromagnetic spectrum
(note: the figures are only approximate)

Radiation	Wavelength (m)	Frequency (Hz)
gamma radiation	$- 10^{-10}$	$10^{19} -$
X-rays	$10^{-12} - 10^{-9}$	$10^{17} - 10^{20}$
ultraviolet radiation	$10^{-9} - 10^{-7}$	$10^{15} - 10^{18}$
visible radiation	$10^{-7} - 10^{-6}$	$10^{14} - 10^{15}$
infrared radiation	$10^{-6} - 10^{-4}$	$10^{12} - 10^{14}$
microwaves	$10^{-4} - 1$	$10^9 - 10^{13}$
radio waves	$1 -$	$- 10^9$

magnetic field in a generator, and by a junction in a thermocouple circuit.

E.m.f. is often loosely used for potential difference; strictly, it should be used only in cases in which a source is driving a current. In a cell, for instance, connected to an external resistance R, the current is given by $I = E/(R + r)$, where E is the e.m.f. and r the source (internal) resistance. The potential difference (V) between the terminals of the cell is ($E - Ir$). This potential difference falls as the current is increased. The e.m.f. in a circuit is the algebraic sum of the potential differences in the circuit ($E = IR + Ir$).

electromotive series (electrochemical series) The arrangement of chemical elements in order of their standard electrode potentials. Elements that tend to lose electrons from their atoms and acquire a positive charge are electropositive. Those that gain electrons are below hydrogen in the table and are electronegative.

A standard hydrogen electrode, at which the reaction $H^+ + e \rightarrow \frac{1}{2}H_2$ takes place, is taken as having zero electrode potential. Elements higher in the series acquire positive charge more easily than those lower down, and tend to displace lower elements from solution. For example, if a strip of zinc is placed in a solution of copper sulphate, copper will be deposited and the zinc will dissolve as ions.

electron An elementary particle of negative charge (-1.602 192 C) and rest mass 9.109 558 kg. Electrons are present in all atoms in shells around the nucleus. They are emitted from the nucleus in a type of beta decay. They are classified as leptons.
See also elementary particles.

electron gun A device that generates a stream of electrons. In a television set, an electron gun produces the moving spot of light on the screen. An electron gun generally consists of a heated cathode in a vacuum tube. The electrons are emitted by thermionic emission and are accelerated towards a positive elec-trode (anode) by a high voltage. They pass through a hole in the anode to form a narrow beam, which can be focused by electron lenses.

electronic energy levels *See* energy levels.

electronics The study and use of circuitry involving such components as semiconductors, thermionic valves and other vacuum devices, resistors, capacitors, and inductors.

electron lens A device for focusing a beam of electrons. Electron lenses use electric fields, produced by a system of metal electrodes, or magnetic fields produced by coils.

electron microscope An apparatus for producing a magnified image by using a beam of electrons focused by electron lenses.

In the *transmission electron microscope* the electron beam passes through the sample and is focused onto a fluorescent screen to produce a visual image. Magnifications above 200 000 can be obtained.

The *scanning electron microscope* can be used with thicker samples. A beam of primary electrons is scanned across the specimen and the secondary electrons emitted are focused onto a screen. The magnification and resolution are lower than in the transmission microscope, but three-dimensional images can be obtained.

In optical microscopes the ultimate resolution depends on the wavelength of the light. Better resolution is obtained using high-energy electrons because their wavelength can be shorter at sufficiently high energies.
See de Broglie wave.

electron optics The use of electric and magnetic fields to focus and direct beams of electrons. Similar methods are used with beams of positive or negative ions.

electron tube An electronic component that depends on flow of electrons in a gas or vacuum; for example, a thermionic valve or discharge tube.

electronvolt Symbol: eV A unit of energy equal to 1.602 191 7 × 10⁻¹⁹ joule. It is defined as the energy required to move an electron charge across a potential difference of one volt. It is normally used only to measure energies of elementary particles, ions, or states.

electrophorus A simple device for supplying electrical charge by electrostatic induction. It consists of a flat insulating plate and a separate metal plate with an insulating handle. The insulating plate is charged by friction; the metal plate is placed on it and momentarily earthed, leaving it with an induced charge.

electroplating The process of coating a solid surface with a layer of metal by means of electrolysis (i.e. by electrodeposition).

electroscope An instrument for detecting electrical potential difference; it can also detect the presence and sign of an electric charge. A common type consists of a pair of light conducting leaves (often of gold foil) suspended side by side from an insulated rod. When a charge is applied to the plate on the top of the rod, the leaves swing apart because they receive electric charges of the same sign.

electrostatic field An electric field produced by stationary (static) electric charge.

electrostatic generator A machine designed to produce a continuous supply of electric charge by electrostatic means. Typical examples are the Wimshurst machine (using friction) and the Van de Graaff generator (using electrostatic induction and discharge from points).

electrostatic induction The production of electric charge at some point by separation of positive and negative charges in an electric field. If a body

with a positive charge (say) is brought close to a neutral conductor, the free electrons in the conductor flow such that they tend to accumulate in the end near the body. This end thus has a negative charge, the other is positive. There is a resulting force of attraction between the body and the conductor. Similarly, an insulator is charged by induction by polarization.

electrostatic precipitation A method of removing unwanted solid or liquid particles from gases (as in pollution control). The particles are given a charge by an electric field, and then attracted to charged plates.

electrostatics (static electricity) The study of the effects associated with electric charge at rest. The fundamental principle of electrostatics is that similarly charged bodies repel each other, whereas oppositely charged bodies attract each other. Electric charges at rest generate electric fields.

electrostatic shield A hollow conducting container surrounding an apparatus, used for protection against electric fields. Faraday demonstrated that the electric field inside a closed conductor is zero. (Electrostatic shields are often called *Faraday cages*.)

electrostatic units *See* c.g.s. system.

electrostriction A change in the dimensions of a dielectric when it is placed in an electric field. It can be regarded as the converse of the piezoelectric effect.

element A single lens forming part of a doublet or other compound lens.

elementary particles (fundamental particles, subatomic particles) Indivisible particles from which all matter is formed. At one time atoms were believed to be the elementary particles. Later it was realized that they were divisible; that is, made of protons, neutrons, and electrons. More recently many new particles have been identified and

Elementary Particles

Particle	Charge	Symbol	Mass	Spin
leptons				
electron	-1	e^-	0.511	1/2
neutrino	0	ν	0	1/2
	0	ν_μ	0	1/2
muon	-1	μ^-	105.66	1/2
baryons				
proton	$+1$	p	938.26	1/2
neutron	0	n	939.55	1/2
xi particle	0	Ξ°	1314.9	1/2
	-1	Ξ^-	1321.3	1/2
sigma particle	$+1$	Σ^+	1189.5	1/2
	0	Σ°	1192.5	1/2
	-1	Σ^-	1197.4	1/2
lambda particle	0	Λ	1115.5	1/2
omega particle	-1	Ω^-	1672.5	3/4
mesons				
kaon	-1	K^-	493.8	0
	$+1$	K^+	493.8	0
pion	$+1$	π^+	139.6	0
	0	π°	135	0
	-1	π^-	139.6	0
phi particle	0	ϕ	1020	1
psi particle	0	ψ	3095	1
eta particle	0	η°	548.8	0

Note: it is common to measure the mass of particles in units of energy/c^2, where c is the speed of light. The values above are in units of MeV/c^2. 1 MeV/$c^2 = 178 \times 10^{-30}$ kg.

elements, magnetic

attempts made to group them into families and explain their relationships with each other.

Elementary particles are characterized by their mass, charge, spin, and certain other properties ('strangeness', 'charm', etc.). One form of classification is into leptons and hadrons, distinguished by the way in which the particles interact. Hadrons are further divided into baryons and mesons. Finally, baryons are subdivided into nucleons and hyperons. Another classification is into *fermions* (having half-integral spin) and *bosons* (having integral spin).

See also antiparticle, quark, resonance.

elements, magnetic *See* magnetic elements.

elliptical polarization A type of polarization of electromagnetic radiation in which the radiation can be considered to consist of two plane-polarized radiations travelling together, out of phase by 90° and of unequal amplitude.

emanation The name formerly given to certain isotopes of radon, which diffused out of the parent material containing radium, from which they are formed by alpha decay. ^{220}Radon is a member of the ^{232}thorium decay series and it was therefore known as *thoron* or thorium emanation. The radon isotope that is a member of the actinium series was known as actinium emanation or *actinon*. Summarizing:

radium emanation is now known as ^{222}Rn

thoron emanation is now known as ^{220}Rn

actinium emanation is now known as ^{219}Rn.

e.m.f. *See* electromotive force.

emission spectrum *See* spectrum.

emissive power *See* emissivity.

emissivity (emissive power) Symbol: ϵ The power radiated by a surface compared to that radiated by a black

body, all other things being equal. There is no unit. Strictly this should be defined for a given wavelength λ (and called *spectral emissivity*). Kirchhoff's radiation law equates the emissivity of a surface to its absorptance. Thus a perfect reflector has zero absorptance and zero emissivity; a black-body has an absorptance of 1 and an emissivity of 1.

emittance *See* exitance.

emitter *See* transistor.

e.m.u. Electromagnetic unit. *See* c.g.s. system.

emulsion A dispersion of droplets of one liquid in another.

endoergic Describing a nuclear process that absorbs energy. *Compare* exoergic.

endothermic Describing a process in which heat is absorbed (i.e. heat flows from outside the system, or the temperature falls). The dissolving of a salt in water, for instance, is often an endothermic process. *Compare* exothermic.

energy Symbol: W A property of a system — its capacity to do work. Energy and work have the same symbol: the joule (J). It is convenient to divide energy into *kinetic energy* (energy of motion) and *potential energy* ('stored' energy). Names are given to many different forms of energy (chemical, electrical, nuclear, etc.); the only real difference lies in the system under discussion. For example, chemical energy is the kinetic and potential energies of electrons in a chemical compound. *See also* internal energy, kinetic energy, mass–energy, potential energy.

energy bands The values of energy that are possible for an electron in a crystalline solid. The electrons in a solid are influenced by the array of positive ions in general, and not just by a single atom. As a result, the discrete energy levels of isolated atoms become bands of allowed

64

energy in solids, separated by gaps of forbidden energy (*forbidden bands*). The valence electrons in the solid are found in an energy band called the *valence band*. The energy band in which electrons are free to move through the solid is the *conduction band*. Within an allowed energy band, the electrons can be in various possible quantum states (one electron in each state).

The band structure of solids explains the differences between conductors, semiconductors, and insulators. For electrons to move through the solid they must be able to change to vacant quantum states at the same energy. In conductors, the valence and conduction bands overlap and there are free states available. In insulators, the filled valence band and empty conduction band are separated by a wide forbidden band, and the electrons thus cannot move. In semiconductors, the forbidden band is narrow and some electrons have enough thermal energy to occupy states in the conduction band.

energy levels One of the discrete energies that an atom or molecule, for instance, can have according to quantum theory. Thus in an atom there are certain definite orbits that the electrons can be in, corresponding to definite *electronic energy levels* of the atom. Similarly, a vibrating or rotating molecule can have discrete vibrational and rotational energy levels.

enrich To increase the fraction of one isotope in a mixture of isotopes of the same element. The term usually refers to the enrichment of uranium by the ^{235}U isotope (natural abundance about 0.7%) to obtain suitable material for nuclear reactors or nuclear weapons. It can also be used to describe the process of obtaining heavy water from natural water. *See also* isotope separation.

enrichment The process of increasing the proportion of one isotope in a given mixture. An example is the enrichment of natural uranium to increase the amount of ^{235}U present, so that the uranium can be used as a fuel in fast

reactors. The term is also applied to the addition of the fissile nuclide ^{239}Pu to uranium for the same purpose.

enthalpy Symbol H The sum of the internal energy (U) and the product of pressure (p) and volume (V) of a system: $H = U + pV$. In a chemical reaction carried out at constant pressure, the change in enthalpy measured is the internal energy change plus the work done by the volume change: $\Delta H = \Delta U + p\Delta V$.

entropy Symbol: S In any system that undergoes a reversible change, the change of entropy is defined as the heat absorbed divided by the thermodynamic temperature: $\delta S = \delta Q/T$. A given system is said to have a certain entropy, although absolute entropies are seldom used: it is change in entropy that is important. The entropy of a system measures the availability of energy to do work.

In any real (irreversible) change in a closed system the entropy increases. Although the total energy of the system has not changed (first law of thermodynamics) the available energy is less — a consequence of the second law of thermodynamics.

The concept of entropy has been widened to take in the general idea of disorder — the higher the entropy, the more disordered the system. For instance, a chemical reaction involving polymerization may well have a decrease in entropy because there is a change to a more ordered system. The 'thermal' definition of entropy is a special case of this idea of disorder — here the entropy measures how the energy transferred is distributed amongst the particles of matter.

epidiascope A form of projector able to project images of both translucent objects (e.g. slides) and flat opaque objects (e.g. diagrams and pictures). It is a combined episcope and diascope. Epidiascopes are now rarely used.

episcope *See* opaque projector.

equation of state An equation relating the pressure, volume, and temperature of a substance. These three quantities define the state of a substance. For an ideal gas the equation of state is:

$$pV = nRT$$

where n is the number of moles of substance and R the molar gas constant. The van der Waals equation is one equation of state for real gases. *See also* gas laws.

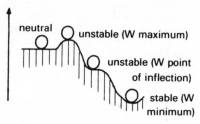

Positions of equilibrium

Quantities used in the equations of motion

Variable	Symbol	Unit symbol
acceleration	a	m/s^2 $(m s^{-2})$
first velocity	v_1 (u)	m/s $(m s^{-1})$
second velocity	v_2 (v)	m/s $(m s^{-1})$
time	t	s
displacement	s	m

Equations of motion

Variables	Equation
$a, v_1, v_2, t, (s)$	$v_2 = v_1 + at^2$
$v_1, v_2, t, s, (a)$	$s = (v_1 + v_2)t/2$
$a, v_1, t, s, (v_2)$	$s = v_1 t + at^2/2$
$a, v_2, t, s, (v_1)$	$s = v_2 t - at^2/2$
$a, v_1, v_2, s, (t)$	$v_2{}^2 = v_1{}^2 + 2 as$

equations of motion Equations that describe the motion of an object with constant acceleration (a). They relate the velocity v_1 of the object at the start of timing to its velocity v_2 at some later time t and the object's displacement s.
The five variables are given in the table. Each of the five equations relates four of the variables, as in the second table, where the omitted variable is shown in brackets.

equator, magnetic *See* isoclinic line.

equilibrant A single force that is able to balance a given set of forces and thus cause equilibrium. It is equal and opposite to the resultant of the given forces.

equilibrium 1. A state of constant momentum. An object is in equilibrium if:
(1) its linear momentum does not change (it moves in a straight line at constant speed and has constant mass, or is at rest);
(2) its angular momentum does not change (its rotation is zero or constant).
For these conditions to be met:
(1) the resultant of all outside forces acting on the object must be zero (or there are no outside forces);
(2) there is no resultant turning-effect (moment).
An object is not in equilibrium if any of the following are true:
(1) its mass is changing;
(2) its speed is changing;
(3) its direction is changing;
(4) its rotational speed is changing.
2. A state of a system in which the properties do not change with time. For example a saturated vapour in contact with its liquid is at equilibrium. The pressure, density, temperature, etc., are unchanging. In this case molecules are still leaving the liquid to the gas phase, but are re-entering from the vapour at the same rate. This type of system is said to be in *dynamic equilibrium*. If two bodies at different temperatures are placed together, heat transfer occurs until they have the same temperature. They are then at *thermal equilibrium*.
A system in which the free energy is at a minimum value is said to be in *thermodynamic equilibrium*. Both the above

66

cases are examples of thermodynamic equilibrium.

equipartition of energy The principle that the total energy of a molecule is, on average, equally distributed among the available degrees of freedom. It is only approximately true in most cases. *See* degrees of freedom.

equipotential A line or surface such that all points on it are at the same potential. Equipotentials in a field cut the lines of force at right angles.

erect Describing an image that is the same way up as the object.

erecting prism (inverting prism) A prism used to invert an image in an optical system without change of size or shape. Erecting prisms operate by internal reflection.

erg A former unit of energy used in the c.g.s. system. It is equal to 10^{-7} joule.

escape velocity The minimum velocity that an object must have in order to escape from the surface of a planet (or moon) against the gravitational attraction. The escape velocity is equal to $\sqrt{(2GM/r)}$, where G is the gravitational constant, M is the mass of the planet, and r is the radius of the planet.

e.s.u. Electrostatic unit. *See* c.g.s. system.

ether (aether) A hypothetical fluid, formerly thought to permeate all space and to be the medium through which electromagnetic waves were propagated.

evaporation 1. A change of state from solid or liquid to gas (or vapour). Evaporation can take place at any temperature; the rate increases with temperature. It occurs because some molecules have enough energy to escape into the gas phase (if they are near the surface and moving in the right direction). Because these are the molecules with higher kinetic energies, evaporation

results in a cooling of the liquid. *See also* vapour pressure.
2. The process of vaporizing a metal at a high temperature in a vacuum. It is used to produce thin metal films.

exchange force A type of force such as that postulated between nucleons in a nucleus. This strong nuclear force operates within a range of up to about 2 $\times$ 10^{-13} cm. The theory holds that protons and neutrons are constantly being transformed into each other, the mediating particle being a meson. *See also* nuclear force.

excitation 1. The production of a magnetic field by an electric current in the winding of an electromagnet.
2. The process of producing an excited state of an atom, molecule, etc.

excitation energy The energy required to change an atom, molecule, etc., from one quantum state to a state with a higher energy. The excitation energy (sometimes called *excitation potential*) is the difference between two energy levels of the system.

excitation potential *See* excitation energy.

excited state A state of an atom molecule, or other system, with an energy greater than that of the ground state. *Compare* ground state.

exclusion principle The principle, enunciated by Pauli in 1925, that no two electrons in an atom could have an identical set of quantum numbers. It has been extended to the nucleons in a nucleus.

exclusive OR gate *See* logic gate.

exitance Symbol: M The (radiant or luminous) flux emitted from a surface per unit area. *Radiant exitance* (M_e) is measured in watts per square metre (W m^{-2}); luminous exitance (M_v) is measured in lumens per square metre (lm m^{-2}). Exitance was formerly called *emittance*.

exoergic Denoting a nuclear process that gives out energy. *Compare* endoergic.

exothermic Denoting a chemical reaction in which heat is evolved (i.e. heat flows from the system or the temperature rises). Combustion is an example of an exothermic process. *Compare* endothermic.

expansion cloud chamber *See* cloud chamber.

expansion, thermal The change of size of a sample (solid, liquid, or gas) when its temperature changes. Normally samples increase in size (expand) when warmed and decrease (contract) when cooled. The effect is due to increased motion (energy) of the atoms or molecules at higher temperatures. (Gases also expand when the pressure is reduced.) *See also* Boyle's law, Charles' law, expansivity.

expansivity A measure of the tendency of a substance to undergo thermal expansion. A solid bar, for example, of length l_1 and temperature θ_1, increases to length l_2 when its temperature rises to θ_2:
$$l_2 = l_1(1 + \alpha(\theta_2 - \theta_1))$$
Here, α is the *linear expansivity* of the material of the bar.

In fact, the expansivity itself varies with temperature, so the mean value of α is obtained between temperatures θ_1 and θ_2. A more accurate equation is:
$$l = l_0(1 + a\theta + b\theta^2 + c\theta^3 + \ldots)$$
l_0 is the length at 0°C, a, b, c, etc., are constants.

Solids also have a *superficial expansivity* (β) relating to the increase in area. It can be shown that, to a good approximation, $\beta = 2\alpha$, and
$$A_2 = A_1(1 + \beta(\theta_2 - \theta_1))$$
Similarly, the cubic expansivity (γ) relates to increase in volume; $\gamma = 3\alpha$:
$$V_2 = V_1(1 + \gamma(\theta_2 - \theta_1))$$
For liquids, only the cubic expansivity (γ) is useful. There are two values

defined. If the expansivity is measured by observing the volume change in a container, the *apparent expansion* is found. This is because the size of the container also changes. The *absolute expansion* is the actual change of volume, allowing for the container. Expansivities were formerly called *coefficients of expansion*. *See also* comparator, Regnault's apparatus.

exposure meter A photocell connected to a suitable display, used to indicate illumination for photography (i.e. to show the correct exposure time and aperture settings).

extraordinary ray *See* birefringent crystal.

extremely high frequency (EHF) A radio frequency in the range between 300 GHz and 30 GHz (wavelength 1 mm–1 cm).

extrinsic semiconductor *See* semiconductor.

eye A sense organ able to produce optical images and to send corresponding nerve signals to the brain. Light enters through the transparent *cornea* and passes through the watery liquid (*aqueous humour*), lens, and gelatinous substance (*vitreous humour*) onto the retina at the back of the eye. The cornea, aqueous humour, and lens form a converging system, which allows an image to be focused on the retina. The focal distance of their system can be varied by the ciliary muscles, which control the curvature of the lens. The quantity of light admitted is varied by the circular iris; this contracts or relaxes, changing the aperture of the eye (the pupil). Light-sensitive cells in the retina produce electrical signals, which are carried along the optic nerve to the brain. The cells are most concentrated at the *fovea* (or *yellow spot*); images produced here are seen in greatest detail.

Two types of cell are found in the retina: the cones, which are used for photopic ('daylight') vision and colour

vision; and rods, which are used for peripheral and scotopic ('night-time') vision.
See also accommodation, colour vision, hyperopia, myopia, presbyopia.

eye lens The lens nearest the eye in a compound eyepiece. *Compare* field lens.

F

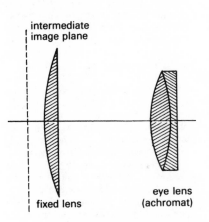

intermediate image plane

fixed lens

eye lens (achromat)

Eye piece (Kellner)

eyepiece (ocular) The lens or combination of lenses nearest the eye in an optical instrument. It is used to produce a final virtual magnified image of the previous image in the system. Various compound lens systems may be used to reduce aberrations. Often crosswires or a graticule will be included in the structure of the eyepiece to assist with location or measurement.

Fahrenheit scale A temperature scale in which the ice temperature is taken as $32°$ and the steam temperature is taken as $212°$ (both at standard pressure). The scale is not used for scientific purposes. To convert between degrees Fahrenheit (F) and degrees Celsius (C) the formula $C/5 = (F - 32)/9$ is used. *See also* temperature scale.

fallout Radioactivity precipitated from the atmosphere, usually originating from nuclear explosions, occasionally from releases of radioactivity from nuclear establishments. It consists of fission fragments, the most worrying of which are ^{131}I (half-life 8 days), which becomes concentrated in the thyroid gland if ingested, and ^{90}Sr (half-life 28 years), which accumulates in bone. Both may reach the human population in milk from cows grazing on land contaminated by fallout and by other less significant routes.

farad Symbol: F The SI unit of capacitance. When the plates of a capacitor are charged by one coulomb and there is a potential difference of one volt between them, then the capacitor has a capacitance of one farad. $1 F = 1 C V^{-1}$. 1 farad = 1 coulomb volt^{-1} $(C V^{-1})$.

faraday Symbol: F A unit of electric charge equal to the charge required to discharge one mole of a singly charged ion. One faraday is $9.648\,670 \times 10^4$ coulombs. *See* Faraday's laws.

Faraday cage *See* electrostatic shield.

Faraday constant Symbol: F The electric charge of one mole of electrons, equal to $9.648\,670 \times 10^4$ coulombs per mole. It is the Avogadro constant multiplied by the electron charge. *See* Faraday's laws.

Faraday disc

Faraday disc *See* homopolar generator.

Faraday effect The rotation of the plane of polarization of radiation in a dielectric medium in a magnetic field. The angle of rotation (θ) is proportional to the field strength B, and to the length l of the path in the medium in the field: θ is proportional to Bl. *See also* Kerr effect.

Faraday's laws (of induction) Three laws describing the electromagnetic induction of an e.m.f. in a conductor by a changing magnetic flux:
(1) If the number of lines of force linked with a conducting circuit is changing, a current is induced in the circuit.
(2) The direction of the induced current is such that the magnetic field it produces tends to keep the flux linkage constant. (This is a statement of Lenz's law).
(3) The total quantity of electricity passing round the circuit is directly proportional to the total change in lines of force (and inversely proportional to the resistance of the circuit).

Faraday's laws (of electrolysis) Two laws resulting from the work of Michael Faraday on electrolysis:
(1) The amount of chemical change produced is proportional to the charge passed.
(2) The amount of chemical change produced by a given charge depends on the ion concerned. More strictly it is proportional to the relative ionic mass of the ion, and the charge on it. Thus the charge Q needed to deposit m grams of an ion, relative ionic mass, M, carrying Z charges, is given by:
$$Q = FmZ/M$$
F, the Faraday constant, has a value of one faraday, i.e. $9.648\,670 \times 10^4$ coulombs.

Faraday's laws of electromagnetic induction *See* electromagnetic induction.

far infrared The longer-wavelength part of the infrared region of the electromagnetic spectrum; i.e. the part furthest in wavelength from the visible region and nearest to the radio-wave region.

far point The furthest point from the eye at which an image can be focused on the retina. The eye is completely relaxed (minimum power) in focusing an object at its far point. In the normal eye the far point is at infinity. In the myopic (short-sighted) eye, light from distant objects is focused in front of the retina. The far point is then only a few metres from the eye. *See also* myopia.

far sight *See* presbyopia.

fast reactor *See* nuclear reactor.

feedback The action of returning some of the output energy of a device (e.g. an amplifier) to the input. There are two types — *positive feedback* and *negative feedback* — according to the relative phases of the feedback voltage and the input signal. The feedback is negative if these voltages are out of phase, and positive if the voltages are in phase.
Although negative feedback reduces the input voltage and input energy, and hence also the gain of the amplifier, the advantage is that noise and distortion are also reduced, while stability is improved. Positive feedback increases the gain and energy output of the amplifier, with the result that the amplifier becomes an oscillator. Feedback is made through a loop circuit using a passive component.
Negative feedback is also used in other control systems. For example, the stabilizer fitted to an ocean-going ship utilizes negative feedback by generating a voltage proportional to the instability (e.g. the angle of tilt) and using this voltage to drive the control system. Thus the effect of the voltage is to reduce the cause which generated it. Another, non-electrical, example is the centrifugal governor.

femto- Symbol: f A prefix denoting 10^{-15}. For example, 1 femtometre (fm) = 10^{-15} metre (m).

Fermat's principle The basic law of geometrical optics. It states that a ray will take a path through a medium such that the time required to cover a given distance is minimum. The law leads to the concept of rectilinear propagation and the laws of reflection and refraction.

fermi A unit of length equal to 10^{-15} metre. It was formerly used in atomic and nuclear physics.

Fermi level The energy level in a solid at which the probability of finding an electron is 1/2. In a conductor the Fermi level lies in the conduction band. At absolute zero all the electrons would occupy energy levels up to the Fermi level, and no higher levels would be occupied. At temperatures above absolute zero, some electrons are in quantum states with energies above this level. In insulators the Fermi level lies in the valence band. In semiconductors it is in the forbidden gap between valence band and conduction band.

fermion See elementary particles.

ferrimagnetism A type of magnetic behaviour shown by certain solids (e.g. ferrites). Ferrimagnetic materials are similar to ferromagnetic materials in behaviour, but have a weaker magnetization. They contain more than one type of magnetic particle with unequal magnetic moments aligned antiparallel, producing a net magnetization. See also antiferromagnetism.

ferrite One of a class of iron compounds with weak permanent magnetism. Ferrites are important as they are electrical insulators so cannot suffer from eddy-current power loss, but they do have a permanent magnetization as a result of their ferrimagnetism. They are used as cores in high-frequency circuits.
  The composition of the ferrites is $Fe_2O_3.MO$, where M is a divalent metal. One common example is magnetite, in which M is Fe^{2+}.

ferromagnetism The magnetism of

substances caused by a domain structure. Such substances have a large susceptibility and relative permeability. Rough values of relative permeability are: iron 1000, cobalt 50, nickel 40. See also Curie's law, hysteresis cycle, magnetism.

fertile material Material that can be converted into fissile material by irradiation with neutrons. ^{238}U and ^{232}Th are fertile materials, being converted into ^{239}Pu (via ^{239}U) and ^{233}U (via ^{233}Th) respectively. Conversion is commonly carried out in breeder reactors. Present breeder reactors use ^{238}U as the fertile material. Breeding from ^{232}Th has not yet been exploited commercially.

FET See field effect transistor.

fibre optics The use of fine transparent fibres to transmit light. The light passes along the fibres by a series of internal reflections. Optical fibres of this type can be used to view inaccessible objects.

field The concept of a field was introduced to explain the interaction of particles or bodies through space. An electric charge, for instance, modifies the space around it such that another charge in this region experiences a force. The region is an electric field (a field of force). Magnetic and gravitational fields can be similarly described.

field coil The winding in a generator or electric motor that produces the magnetic field.

field effect transistor (FET) A solid-state electronic device with three terminals that, like the junction transistor, is used in amplifiers. It controls the current between two terminals, the source and the drain, by the voltage at a third, the gate. An n-type FET consists of a single piece of n-type semiconductor, which has the source at one end and the drain at the other. A heavily-doped p-type region in the middle forms the gate. If the gate voltage is more negative than the source, electrons move from the n-type region to the p-type region leaving an area round the gate with fewer electrons to carry current. This has the effect of narrowing the conducting channel through which the source–drain current flows by an amount that depends on the source-gate voltage difference. In the FET, only one type of charge carrier — electrons in the n-type FET and holes in the p-type FET — determines the current and it is therefore known as a *unipolar* transistor. In the *bipolar* junction transistor, both positive and negative charge carriers contribute to the other current. *See also* transistor.

field emission (cold emission) The release of electrons from a surface as a result of a strong external electric field. Very high electric fields are necessary; these are obtained at sharp points.

field lens The lens furthest from the eye in a compound eyepiece. *Compare* eye lens.

field strength, electric *See* electric field strength.

filament A fine thread of metal, glass, etc.

filter 1. A device that passes only certain frequencies of radiation. A red filter, for example, is a piece of translucent material that transmits red light, absorbing other wavelengths.
2. An electronic circuit that transmits currents within a certain range of frequencies, while reducing (attenuating) other frequencies.

fine structure Closely spaced lines seen at high resolution in a spectral line or band. Fine structure may be caused by vibrational motion of the molecules or by electron spin. *Hyperfine structure*, seen at very high resolution, is caused by the atomic nucleus affecting the possible energy levels of the atom.

fissile material Material that under-

goes nuclear fission, sometimes spontaneously but usually when irradiated by neutrons. Examples are ^{235}U, ^{239}Pu, and ^{233}U. Fissile material is the fuel of nuclear reactors and the explosive material of nuclear weapons.

fission *See* nuclear fission.

fission reactor *See* nuclear reactor.

fission-track dating A method of measuring the age of glasses and other minerals that depends on the tracks made in these solids by fragments from the spontaneous fission of contained uranium. The density of tracks in the material depends on the uranium content, the age of the material, and any fading of the tracks. After counting the tracks, the uranium content is estimated by irradiating the materials with neutrons to induce fissions. The number of extra tracks produced depends on the uranium content, which can then be estimated, giving the age since solidification.

fixed point *See* temperature scale, International Practical Temperature Scale.

Fleming's rules A method of remembering the relationship between the directions of current, magnetic field, and motion in electric motors and generators. Fleming's *left-hand rule* applies to motors. If the left hand is held with the thumb and first two fingers all mutually at right angles, the thumb shows *motion* (force); the *first* finger, *field*; and the second finger, *current*. Fleming's *right-hand rule* is similar but with the fingers of the right hand showing the operation of an electrical generator.

flint glass *See* optical glass.

flip-flop *See* bistable circuit.

floating objects, law of *See* flotation, law of.

flotation, law of An object floating in a fluid displaces its own weight of fluid.

This follows from Archimedes' principle for the special case of floating objects. (A floating object is in equilibrium, its only support coming from the fluid. It may be totally or partly submerged.)

fluid A substance that flows; i.e. a liquid or a gas.

fluidics The study and use of fluid flow through pipes in an analogous way to the flow of electric current through circuits. Fluidic circuits may be useful in circumstances in which there are high magnetic fields or radiation levels, which would affect electrical circuitry.

fluorescence The absorption of energy by atoms, molecules, etc., followed by immediate emission of electromagnetic radiation as the particles make transitions to lower energy states. Fluorescence is a type of luminescence in which the emission of radiation does not persist after the exciting source has been removed. The excited states have very short lifetimes.

Fluorescence, like luminescence, can be produced in a number of ways. Examples are by bombardment with electrons and by absorption of other electromagnetic radiation. In the case of electromagnetic radiation the emitted radiation often has lower frequency than the absorbed radiation. A particular example is the absorption of ultraviolet radiation followed by emission of visible radiation. This effect is used in fluorescent paints and in 'fabric brighteners' added to detergents. The inside coating of fluorescent tubes is another example of a material that converts ultraviolet radiation (from the discharge) into visible radiation. *See also* luminescence.

fluorescent lamp A lamp in which light is generated by an electrical discharge through a low-pressure gas, causing it to emit radiation. The radiation produced is usually ultraviolet radiation, and is changed into visible light when it strikes the inner surface of the gas-discharge tube, which is coated with a phosphor.

fluoroscope An instrument that makes X-rays, or other non-visible radiation, visible by means of a fluorescent screen.

flux In general, a flow or apparent flow:
1. A flow of particles per unit area.
2. The density of lines of force in a field. *See* magnetic flux.
3. *See* luminous flux.

flux density *See* magnetic flux density.

fluxmeter An instrument for measuring magnetic flux. The most common type uses a small movable exploring coil (a *search coil*), which is placed in the field and then removed quickly. The induced current can be measured by a ballistic galvanometer. *See also* Earth inductor.

flywheel A large heavy wheel (with a large moment of inertia) used in mechanical devices. Energy is used to make the wheel rotate at high speed; the inertia of the wheel keeps the device moving at constant speed, even though there may be fluctuations in the torque. A flywheel acts as an 'energy-storage' device.

FM *See* frequency modulation.

f-number For an optical instrument, especially a camera, the ratio of the focal distance (f) to the aperture diameter (d):
$$\text{f-number} = f/d$$
If the ratio is 16 the f-number is written f16, f:16, or f/16.

focal distance (focal length) A measure of the power of a lens or mirror to converge a parallel beam. The focal distance of a lens or mirror is the distance between the pole and the focal point. Normally, positive values relate to converging lenses and mirrors, and negative values relate to diverging ones, (real is positive, virtual is negative convention). The New Cartesian convention is less common. It gives positive focal distances to diverging mirrors and converging lenses, and negative values for converging mirrors and diverging lenses.

For a refracting surface:
$$f = n_1 r/(n_2 - n_1)$$
where n is refractive constant and r is radius of curvature.

For a reflecting surface:
$$f = r/2$$

For a thin lens in air:
$$1/f = (n - 1)(1/r_1 - 1/r_2)$$

For two thin lenses in contact:
$$1/f = 1/f_1 + 1/f_2$$

focal length *See* focal distance.

focal plane The plane perpendicular to the axis centred on the focal point of a lens or mirror.

focal point (focus, principal focus) A point through which rays close and parallel to the axis pass or appear to pass after reflection or refraction. A reflector has one focal point. As radiation can enter a lens from either side, a lens has a focal point on each side, at the same distance from the lens.

Only rays close and parallel to the axis pass through the focal point. These are called *paraxial* rays. Rays parallel to the axis but not close to it focus somewhere on the caustic surface. Rays close to the axis but not parallel to it focus somewhere on the focal plane.

focal power *See* power.

focus 1. *See* focal point.
2. An image is in focus if it is sharp rather than blurred (out of focus). Only then do the set of rays from a point on the object all pass through a single point.
3. A point through which each member of a set of rays passes or appears to pass after reflection or refraction. Paraxial rays focus at the focal point; other sets may focus on the focal plane or on the caustic surface. Others again, from the surface of an object, focus on the image plane to form the image.

foot Symbol: ft The unit of length in the

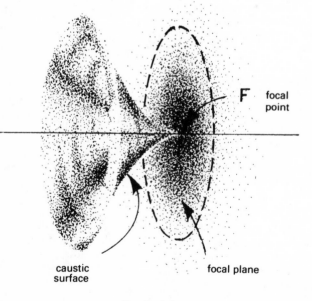

caustic surface   focal plane

Focal point

Focal points for lenses and mirrors

	Focal point	Position	Focal distance
converging lens	real	far side of lens	positive
diverging lens	virtual	near side of lens	negative
converging mirror	real	in front of mirror	positive
diverging mirror	virtual	behind mirror	negative

f.p.s. system (one third of a yard). It is equal to 0.304 8 metre.

forbidden band *See* energy bands.

force Symbol: *F* That which tends to change an object's momentum. Force is a vector; the unit is the newton (N).
In SI, this unit is so defined that:
$$F = \mathrm{d}(mv)/\mathrm{d}t$$
(from Newton's second law). *See* Newton's laws of motion.

forced convection *See* convection.

forced oscillation (forced vibration) The oscillation of a system or object at a frequency other than its natural frequency. Forced oscillation must be induced by an external periodic force. *Compare* free oscillation. *See also* resonance.

force ratio (mechanical advantage) For a machine, the ratio of the output force (load) to the input force (effort). There is no unit; the ratio is, however, often given as a percentage. It is quite possible for force ratios far greater than one to be obtained. Indeed the design of many machines is to make this so, to make a small effort overcome a large load.
However efficiency cannot be greater than one; a large force ratio implies a large distance ratio. *See also* machine.

forces, parallelogram (law) of *See* parallelogram of vectors.

forces, triangle (law) of *See* triangle of vectors.

formula 1. A general rule, often represen-

ted by symbols. For example, $A = \pi r^2$, is a formula giving the area of a circle in terms of the radius.
2. A representation of a compound (or molecule) using the symbols for the elements.

Foucault pendulum A simple pendulum consisting of a heavy bob on a long string. The plane of vibration rotates slowly over a period of time as a result of the rotation of the Earth below it.
The apparent force causing this movement is the Coriolis force.

fovea (yellow spot) A small area of the retina where the cones are most dense. It is the part with the greatest visual acuity. *See* eye.

frame of reference A set of coordinate axes by which the position of any object may be specified as it changes with time. The origin of the axes and their spatial direction must be specified at every instant of time for the frame to be fully determined.

Fraunhofer diffraction Diffraction observed with incident parallel light. In Fraunhofer diffraction the wavefronts are parallel. Although a special case of Fresnel diffraction, it is far more important in most practical cases. Thus it is used to explain single — and multiple — slit patterns, as well as those produced by circular holes.

Fraunhofer lines The dark lines in the spectrum of light from the Sun, caused by the absorption of particular wavelengths by certain elements in its cooler outer regions. The wavelengths of these lines are used as reference points in specifying quantities that vary with wavelength, e.g. refractive constants.

free energy A measure of the ability of a system to do useful work. *See* Gibbs function; Helmholtz function.

free oscillation (free vibration) An oscillation at the natural frequency of the system or object. Thus a pendulum can be forced to swing at any frequency; it will swing freely at only one, which depends on its length and mass. *Compare* forced oscillation. *See also* resonance.

free surface energy *See* surface tension.

free fibration *See* free oscillation.

freezing The change from the liquid to the solid state that occurs when energy is transferred from a substance. For pure substances this happens at a characteristic temperature known as the *freezing point* (or *melting point*). This depends on pressure, so is usually measured at standard atmospheric pressure. Impurities generally lower the freezing point.

freezing mixtures Two or more substances mixed together to produce a low temperature. A mixture of salt and ice in water is a common example.

freezing point *See* freezing.

Frenkel defect *See* defect.

frequency Symbol: f, ν The number of cycles per unit time of an oscillation (e.g. a pendulum, vibrating system, wave, alternating current, etc.). The unit is the hertz (Hz).

Angular frequency (and pulsatance) are related to frequency by $\omega = 2\pi f$.

frequency modulation (FM) A type of modulation in which a carrier wave is made to carry the information in a signal (audio or visual) by fluctuations in the frequency of the carrier. The variation of the carrier frequency is proportional to the frequency of the signal, while the amplitude of the carrier remains constant. Frequency modulation is superior to amplitude modulation because a wider band of frequencies may be transmitted with less interference and noise. *See also* carrier wave, modulation.

Fresnel biprism *See* biprism, Fresnel's.

Fresnel diffraction Diffraction observed when either source or screen (or both) are close to the diffractor. In Fresnel diffraction, the wavefronts are not plane (as in Fraunhofer diffraction) and analysis is difficult. The approach is useful in explaining (for instance) diffraction around a circular obstacle.

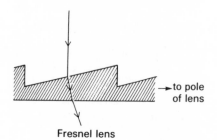

Fresnel lens

Fresnel lens A type of lens with one surface cut in steps so that transmitted light is refracted just as if by a much thicker (and heavier and more expensive) conventional lens. Light-house lamp lenses have long been of this type. Very cheap plastic Fresnel lenses now have many uses, for example in overhead projectors, flashlamps, etc. The angle of each step is cut to produce the desired effect.

Fresnel zones *See* half-period zones.

friction A force opposing the relative motion of two surfaces in contact. In fact, each surface applies a force on the other in the opposite direction to the relative motion; the forces are parallel

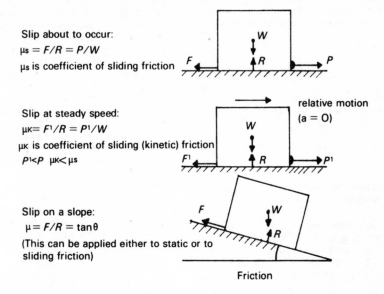

Slip about to occur:
$$\mu_s = F/R = P/W$$
μ_s is coefficient of sliding friction

Slip at steady speed:
$$\mu_K = F^1/R = P^1/W$$
μ_K is coefficient of sliding (kinetic) friction
$$P^1 < P \quad \mu_K < \mu_s$$

relative motion
$(a = 0)$

Slip on a slope:
$$\mu = F/R = \tan\theta$$
(This can be applied either to static or to sliding friction)

Friction

to the line of contact. The exact causes of friction are still not fully understood. It probably results from minute surface roughness, even on apparently 'smooth' surfaces. Frictional forces do not depend on the area of contact. Presumably lubricants act by separating the surfaces.

For friction between two solid surfaces, *sliding* (or *kinetic*) friction opposes friction between two moving surfaces. It is less than the force of *static* (or *limiting*) *friction*, which opposes slip between surfaces that are at rest. *Rolling friction* occurs when a body is rolling on a surface: here the surface in contact is constantly changing. Frictional force (F) is proportional to the force holding the bodies together (the 'normal reaction' R). The constants of proportionality (for different cases) are called *coefficients of friction* (Symbol: μ): $\mu = F/R$.

The *laws of friction* are sometimes stated:
(1) The frictional force is independent of the area of contact (for the same force holding the surfaces together).
(2) The frictional force is proportional to the force holding the surfaces

together. In sliding friction it is independent of the relative velocities of the surfaces.
Frictional forces also occur in fluids. *See* viscosity.

fringes The light and dark bands obtained by interference or diffraction of light.

fuel cell A type of cell in which fuel is converted directly into electricity. †In one form, hydrogen gas and oxygen gas are fed to the surfaces of two porous nickel electrodes immersed in potassium hydroxide solution. The oxygen reacts to form hydroxyl (OH^-) ions, which it releases into the solution, leaving a positive charge on the electrode. The hydrogen reacts with the OH^- ions in the solution to form water, giving up electrons to leave a negative charge on the other electrode. Large fuel cells can generate tens of amperes. Usually the e.m.f. is about 0.9 volt and the efficiency around 60%.

fundamental The simplest way (mode) in which an object can vibrate. The fundamental frequency is the frequency of this vibration. The less simple modes of vibration are the higher *harmonics*; their frequencies are higher than that of the fundamental. *See also* quality of sound.

fundamental constants (universal constants) Quantities that do not change under any known circumstances. Examples are the speed of light in a vacuum and the electric charge on an electron.

fundamental interval *See* temperature scale.

fundamental particles *See* elementary particles.

fundamental units The units of length, mass, and time that form the basis of most systems of units. In SI, the fundamental units are the metre, the kilogram, and the second. *See also* base unit.

fusion 1. The combination of two nuclei into a heavier nucleus. The process should not be confused with a chemical reaction, which involves only the orbiting electrons. In the case of light atomic nuclei, nuclear fusion is associated with the production of large amounts of free energy. Before such reactions can occur energy has to be supplied to overcome the repulsive force between the nuclei resulting from their electric charge. Fusion reactions thus need a high initiating temperature, and are called *thermonuclear reactions*. In thermonuclear weapons (e.g. the hydrogen bomb) the reaction is initiated by a fission reaction. Controlled fusion reactions for producing usable energy are the subject of much research. There are problems in initiating the reaction and containing the plasma. Fusion reactions are thought to be responsible for the high temperatures of stars. *See also* carbon cycle, proton–proton chain reaction.
2. *See* melting.

fusion reactor *See* nuclear reactor.

Some physical constants

Quantity	Magnitude	Unit
speed of light	$2.997\,925 \times 10^8$	$m\,s^{-1}$
Planck constant	$6.626\,196 \times 10^{-34}$	$J\,s$
Boltzmann constant	$1.380\,622 \times 10^{-23}$	$J\,K^{-1}$
Avogadro constant	$6.022\,169 \times 10^{23}$	mol^{-1}
mass of proton	$1.672\,614 \times 10^{-27}$	kg
mass of neutron	$1.674\,920 \times 10^{-27}$	kg
mass of electron	$9.109\,558 \times 10^{-31}$	kg
charge of proton or electron	$\pm 1.602\,191\,7 \times 10^{-19}$	C
specific charge of electron	$-1.758\,796 \times 10^{11}$	$C\,kg^{-1}$
molar volume at s.t.p.	$2.241\,36 \times 10^{-2}$	$m^3\,mol^{-1}$
Faraday constant	$9.648\,670 \times 10^4$	$C\,mol^{-1}$
triple point of water	273.16	K
absolute zero	-273.15	$°C$
permittivity of vacuum	$8.854\,185\,3 \times 10^{-12}$	$F\,m^{-1}$
permeability of vacuum	$4\pi \times 10^{-7}$	$H\,m^{-1}$
Stefan constant	$5.669\,61 \times 10^{-8}$	$W\,m^{-2}K^{-4}$
molar gas constant	$8.314\,34$	$J\,mol^{-1}K^{-1}$
gravitational constant	$6.673\,2 \times 10^{-11}$	$N\,m^2\,kg^{-2}$

G

gain A ratio measuring the efficacy of an electronic system, usually an amplifier or an aerial. The voltage gain of an amplifier is the ratio of the output voltage developed across a load to the input voltage. The gain of a power amplifier is the ratio of the output power to the input power. The gain of a directional aerial is the ratio of the voltage generated when the aerial is orientated with optimum effect to the voltage generated by the same incoming signal in an aerial without directional properties.

Galilean telescope A type of refracting telescope having a converging objective and a diverging eyepiece. The Galilean arrangement produces an upright final image. However the field of view is smaller than that of the Keplerian telescope (which gives an inverted image). The arrangement is used in cheap binoculars (opera glasses). The normal adjustment of this telescope has the intermediate image in the focal planes of both lenses. These are therefore separated by a distance $f_O - f_E$.
See also refractor.

galvanic cell *See* cell.

galvanometer An instrument for measuring small direct currents. *See* moving-coil instrument, tangent galvanometer. *See also* ballistic galvanometer.

gamma (γ) The ratio of the principal specific thermal capacity of a gas at constant pressure to that at constant volume (c_p/c_v). *See* adiabatic change, sound, specific thermal capacity.

gamma decay A type of radioactive decay in which gamma rays are emitted by the specimen. Gamma decay occurs when a nuclide is produced in an excited state, gamma emission occurring by transition to a lower energy state. It can occur in association with alpha decay and beta decay.

gamma radiation A form of electromagnetic radiation emitted by changes in the nuclei of atoms. Gamma waves have very high frequency (short wavelength); they carry much energy, so can travel a long way through matter before being totally absorbed. Gamma radiation absorbed by living cells can cause damage.

Gamma radiation shows particle properties much more often than wave properties. The energy of a gamma photon, given by $W = h\nu$, can be very high. (h is the Planck constant) A gamma photon of 10^{24} Hz has an energy of 6.6×10^{-10} J, an enormous amount for such a small particle. The energy of low-energy gamma rays can be measured

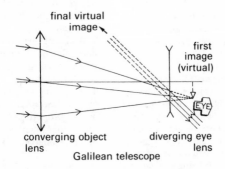

final virtual image

first image (virtual)

EYE

converging object lens

diverging eye lens

Galilean telescope

by diffraction (a wave property enabling wavelength to be found). High-energy photons can be measured by the ionization produced in radiation detectors. *See also* electromagnetic spectrum.

gamma rays Streams of gamma radiation.

gas *See* states of matter.

gas constant (universal molar gas constant) Symbol: R The universal constant 8.314 34 J mol^{-1} K^{-1} appearing in the equation of state for an ideal gas.

gas-cooled reactor A nuclear reactor in which cooling of the core is achieved by circulation of a gas. Early reactors were cooled by air. Magnox reactors are cooled by high-pressure carbon dioxide. The advanced *gas-cooled reactors* (*AGR*) will also be cooled by carbon dioxide. The *high-temperature gas-cooled reactor* (*HTGR*) is cooled by helium.

gas laws Laws relating the temperature, pressure, and volume of a fixed mass of gas. The main gas laws are Boyle's law and Charles' law. The laws are not obeyed exactly by any real gas, but many common gases obey them under certain conditions, particularly at high temperatures and low pressures. A gas that would obey the laws over all pressures and temperatures is a *perfect* or *ideal gas*.

Boyle's and Charles' laws can be combined into an equation of state for ideal gases:
$$pV_m = RT$$
where V_m is the molar volume and R the molar gas constant. For n moles of gas
$$pV_m = nRT$$
All real gases deviate to some extent from the gas laws, which are applicable only to idealized systems of particles of negligible volume with no intermolecular forces. There are several modified equations of state that give a better description of the behaviour of real gases, the best known being the van der Waals equation. Deviations from the ideal-gas laws can be demonstrated by plotting pV against p for various temperatures. For an ideal gas the graph is a line parallel to the pressure axis (pV is a constant). For real gases, pV varies with pressure, although there is one particular temperature (the *Boyle temperature*) at which the gas obeys Boyle's law.

gas thermometer A thermometer that uses a fixed mass of gas as its working substance. It can take two forms. A *constant volume gas thermometer* depends on the direct proportion relationship between the pressure of a fixed mass of gas and its absolute temperature at constant volume. A *constant pressure gas thermometer* is dependent on the similar relationship between the volume and temperature of the gas.

Constant volume gas thermometers may use helium, hydrogen, or nitrogen. The typical arrangement is a bulb connected to a U-tube containing mercury. As the temperature changes the heights of mercury are adjusted such that the volume of gas in the bulb is constant. The pressure is measured by differences in the heights of the mercury. Corrections can be made for expansion of the bulb, temperature of the mercury, and nonideal behaviour of the gas.

gate 1. A signal that enables a circuit to function.
2. An electrode in a semiconductor device to which a biasing voltage is applied to modify the conductivity. *See* logic gate.
3. A circuit that has one or more inputs but only one output. This is a digital gate as used in logic circuits. *See* field effect transistor.
4. A circuit that admits an input for only a specified time, producing an output proportional to the input, or to some other function of the input. *See also* semiconductor.

gauss Symbol: G The unit of magnetic

flux density in the c.g.s. system. It is equal to 10^{-4} tesla.

Gay-Lussac's law *See* Charles' law.

Geiger counter (Geiger–Muller counter) A device for detecting ionizing radiation to provide a count of individual particles and photons. The construction is a gas-filled tube with a cylindrical metal cathode and a thin axial wire anode. The potential difference applied between the two depends on the design, but is usually between about 400 and 2000 volts. The gas may be low-pressure air or argon with ethanol vapour. Low-energy particles can enter through a thin 'window', often of mica. Passage of an alpha or beta particle or a photon of gamma radiation through the gas produces ions, which move towards the electrodes. The electrons accelerating towards the central anode acquire sufficient kinetic energy to ionize other gas molecules, initiating an 'avalanche' of electrons, which produces a measurable pulse at the anode. The pulse of current can be used to operate counting equipment. Some Geiger tubes contain a small amount of halogen vapour to quench the avalanche rapidly so that the tube returns quickly to a state in which it is ready to detect another pulse. Several thousand particles per second can be counted in this way.

General theory *See* relativity.

generator A large machine for converting mechanical energy into electrical energy. It has two sets of coil windings. One (the stator) is fixed and the other (the rotor) rotates. The relative motion of the two sets of coils induces an electric current in them. This is the reverse of a motor. In an electrical power station, generators are usually turned by steam turbines or water turbines. The turbine turns a generator, often called an alternator, that produces alternating current of a fixed frequency. Diesel engines are used to turn smaller generators of the type kept for emer-gency use in hospitals, etc. *See also* dynamo, electric motor.

geomagnetism *See* Earth's magnetism.

geometrical optics *See* optics.

getter A chemical substance used for removing residual gases after a vacuum has been produced by pumping. Getters are usually metals that combine with oxygen or nitrogen to produce a very high vacuum.

GeV Gigaelectronvolts; 10^9 electronvolts. Often *BeV* is used in the U.S.

ghost A faint image near the image required, caused by radiation that has taken a different path. In the case of traditional 'silvered' glass mirrors, the main images are caused by light from the backing of the glass. Ghost images, formed by reflection from the front surface of the glass, may be a tenth as bright as the main images. Other 'ghost' images may arise following reflections inside the glass.

Ghost effects can also arise with non-visible electromagnetic radiations. Thus, ghost television images can appear on the screen if some radio waves reach the aerial after reflection by large nearby objects. This effect can be used to obtain a fairly accurate value for the speed of radio waves.

Gibbs function (Gibbs free energy) Symbol: G A thermodynamic function defined by $G = H - TS$, where H is the enthalpy, T the thermodynamic temperature, and S the entropy. It is useful for specifying the conditions of chemical equilibrium for reactions for constant temperature and pressure (G is a minimum). *See also* free energy.

giga- Symbol: G A prefix denoting 10^9. For example, 1 gigahertz (GHz) = 10^9 hertz (Hz).

glass A solid substance in which there is a non-regular arrangement of atoms. Glasses are thus not crystalline — their atoms have a fairly random arrangement and they soften over a range of temperature, showing no definite melting point. They can be regarded as supercooled liquids. *See also* optical glass.

glow discharge A silent discharge of electricity through a gas at low pressure. Usually luminous, it consists of dark and light bands between the electrodes.

gold-leaf electroscope *See* electroscope.

governor A mechanical device to control the speed of rotation of a machine. A simple governor consists of two weights attached to a shaft so that as the speed of rotation of the shaft increases, the weights move further outwards from the centre of rotation, while still remaining attached to the shaft. As the weights move outwards they operate a control that reduces the fuel or energy input to the machine. As they reduce speed and move inwards they increase the fuel or energy input. Thus, on the principle of negative feedback, the speed of the machine is kept constant under varying conditions of load.

grade A unit of angle equal to $1/100$ of a right angle. $1^g = 0.9°$.

Graham's law (of diffusion) Gases diffuse at a rate that is inversely proportional to the square root of their density. Light molecules diffuse faster than heavy molecules. The principle is used in the separation of isotopes.

gram Symbol: g A unit of mass defined as 10^{-3} kilogram.

gram-atom *See* mole.

gram-molecule *See* mole.

graticule A grid of squares used with the eyepieces of many optical instruments to provide easier measurement and/or reference.

grating *See* diffraction grating.

gravitation The concept originated by Isaac Newton around 1666 to account for the apparent motion of the Moon about the Earth, the essence being a force of attraction, called gravity, between the Moon and the Earth. Newton used this theory of gravitation to give the first satisfactory explanations of many diverse physical facts, such as Kepler's laws of planetary motion, the ocean tides, and the precession of the equinoxes. *See also* Newton's law of universal gravitation.

gravitational constant Symbol: G The constant of proportionality in the equation that expresses Newton's law of universal gravitation: $F = Gm_1m_2/r^2$, where F is the gravitational attraction between two point masses, m_1 and m_2, separated by a distance r. The value of G is 6.67×10^{-11} $Nm^2 kg^{-2}$. It is regarded as a universal constant, although it has been suggested that the value of G may be changing slowly owing to the expansion of the Universe. *See also* Newton's law of universal gravitation.

gravitational field The region of space in which one body attracts other bodies as a result of their mass. Bodies on or near the surface of the Earth are influenced by the Earth's gravitational field. To escape from this field a body has to be projected outwards with a certain velocity (the *escape velocity*). The strength of the gravitational field at a point on the Earth's surface is given by the ratio force/mass, which is equivalent to the acceleration of free fall. At a point, it is defined as GM/r^2, where G is the gravitational constant, M the mass of the Earth, and r the distance between the centre of the Earth and the point in question. The standard value of the acceleration of free fall is 9.8 m s^{-2}, but it varies with altitude (i.e. with r^2).

gravitational interaction *See* interaction.

gravitational mass The mass of a body as measured by the force of attraction between masses, the value being given by Newton's law of universal gravitation. Inertial and gravitational masses are equal in a uniform gravitational field. *See also* inertial mass.

gravitational red shift *See* red shift.

graviton A fundamental particle postulated in order to explain the very weak gravitational interaction in a way that fits in with quantum mechanics. Gravitons have zero rest mass and travel at the speed of light between the two interacting masses. None have yet been observed.

gravity The gravitational pull of the Earth (or other celestial body) on an object. The force of gravity is identical to weight.

gravity, centre of *See* centre of mass.

gray Symbol: Gy The SI unit of absorbed energy dose per unit mass resulting from the passage of ionizing radiation through living tissue. One gray is an energy absorption of one joule per kilogram of mass. *See also* dose.

Gregorian telescope An early type of reflecting telescope similar to the Cassegrainian arrangement, except that both mirrors were converging. *See also* reflector.

grid 1. An open wire mesh interposed between the anode and cathode of a thermionic valve, the purpose being to modify the current–voltage characteristic, and in particular to make amplification possible. *See also* thermionic valve.
2. The system of distributing electrical power nationally at high voltage (up to 400 kV).

grid bias The steady potential, usually negative, applied to the control grid of a thermionic valve. The grid bias has a value such that, when a fluctuating input signal voltage is superimposed, the valve is neither cut off nor made

conducting through the grid. *See also* grid, thermionic valve.

grounding *See* earthing.

ground state The lowest energy state of an atom, molecule, or other system. *Compare* excited state.

ground waves A radio wave that travels close to the surface of the Earth between the transmitter and the receiver. *Compare* sky wave.

group velocity If a wave motion has a phase velocity that depends on wavelength, the disturbance of a progressive wave travels with a different velocity from the phase velocity. This is called the *group velocity*. The group velocity is the velocity with which the group of waves travels. It is given by $U = c - \lambda dc/d\lambda$, where c is the phase velocity. The group velocity is the one that is usually obtained by measurement. If there is no dispersion, as for electromagnetic radiation in free space, the group and phase velocities are equal.

gyroscope A rotating object that tends to maintain a fixed orientation in space. For example, the axis of the rotating Earth always points in the same direction towards the Pole Star (except for a small precession). A spinning top or a cyclist are stable when moving at speed because of the gyroscopic effect. Practical applications are the navigational gyrocompass and automatic stabilizers in ships and aircraft. *See also* precession.

H

hadrons A group of elementary particles, subdivided into the baryons and the mesons. The hadrons are distinguished from the leptons by their type of interaction. *See also* elementary particles.

Hagen's formula *See* Poiseuille's formula.

half cell A metal electrode in contact with a solution of the metal ions. In general there will be an e.m.f. set up between metal and solution, depending on the tendency of the element to form ions in solution. The e.m.f. cannot be measured directly since setting up a circuit results in the formation of another half cell. *See* electrode potential.

half-life (half-life period) Symbol: $T_{\frac{1}{2}}$ The time taken for half the nuclei of a sample of a radioactive nuclide to decay. The half-life of a nuclide is a measure of its stability. (Stable nuclei can be thought of as having infinitely long half-lives.) Half-lives of known nuclear species range from less than a picosecond to thousands of millions of years. If N_0 is the original number of nuclei, the number remaining at the end of one half-life is $N_0/2$, at the end of two half-lives is $N_0/4$, etc. *See also* decay.

half-period zones (Fresnel zones) The circular bands on a spherical wavefront required in the analysis of Fresnel diffraction. Each band is half a wavelength closer to, or further from, the point of observation than its neighbour. Their radii are $\sqrt{(nd\lambda)}$, where n (the order of the zone) is an integer, and d is the distance from the observation point to the centre of the wavefront.

half-wave plate *See* retardation plate.

half-width *See* monochromatic radiation.

Hall effect When an electric current is passed through a conductor and a magnetic field is applied at right angles, a potential difference is produced between two opposite surfaces of the conductor. The direction of the potential gradient is perpendicular to both the current direction and the field direction. It is caused by deflection of the moving charge carriers in the magnetic field. The size and direction of the potential difference gives information on the number and type of charge carriers.

hard (high) vacuum *See* vacuum.

harmonic One of the possible simple (sinusoidal) components of a complex waveform or vibration. The *first harmonic* is the fundamental (frequency f); the *second harmonic* has twice the fundamental frequency ($2f$), and so on. A waveform may contain other frequencies than these harmonics. *See also* partials, quality of sound.

harmonic motion A regularly repeated sequence that can be expressed as the sum of a set of sine waves. Each component sine wave represents a possible simple harmonic motion. The complex vibration of sound sources (with fundamental and overtones), for instance, is a harmonic motion, as is the sound wave produced. *See also* simple harmonic motion.

heat Symbol: Q The name often given to energy that is transferred from regions of high temperature to those of lower temperature. The energy transferred when a system changes temperature is equal to the product of the temperature change, the mass of the object, and its specific thermal capacity. The energy transferred when a sample changes state is the product of mass and specific latent thermal capacity. *See also* internal energy, kinetic theory.

heat capacity *See* thermal capacity.

heat engine (thermodynamic engine) A device for converting heat into work. Heat engines operate by transferring energy from a high temperature to a low temperature sink. The theoretical operation of heat engines is useful in thermodynamics. *See* Carnot cycle.

heat exchanger A device for transferring heat from one fluid to another. Typically, a heat exchanger consists of a series of tubes through which one fluid flows. These are surrounded by the other hotter or cooler fluid. A car radiator is an example of a heat exchanger for transferring energy from the engine coolant to the surroundings.

heat pipe A device for conducting energy. It consists of a sealed tube containing a volatile liquid under low pressure. When one end of the tube is heated, the liquid vaporizes, and molecules carry energy to the other end of the tube, where they condense. Liquid returns to the higher-temperature end of the pipe by capillary action through a wire-mesh coating on the inside of the pipe. Heat pipes are very efficient; a pipe can conduct energy 1000 times as quickly as a piece of copper with the same dimensions.

heat pump A device that transfers energy from a low temperature region to a high temperature region by doing work. For example, a refrigerator removes heat from the cold interior of a box and transfers it to the higher temperature surroundings.

heavy hydrogen *See* deuterium.

heavy water *See* deuterium.

hecto- Symbol: h A prefix denoting 10^2. For example, 1 hectometre (hm) = 10^2 metres (m).

Heisenberg's uncertainty principle
The impossibility of making simultaneous measurements of both the position and the momentum of a subatomic particle such as an electron, with unlimited accuracy. This is because, in order to detect the particle, radiation has to be 'bounced' off it, and this process itself disrupts the particle's position. Heisenberg's uncertainty principle represents a fundamental limit to 'objective' scientific observation. It arises from the wave-particle duality of particles and radiation. In one direction the uncertainty in position Δx and

momentum Δp_x are related by $\Delta x \Delta p \sim h/4\pi$, where h is the Planck constant.

Helmholtz coils A pair of identical coils designed so that when a current is carried the magnetic field between them is fairly uniform. The two coils are set parallel at a distance equal to the radius of each. They are connected in series so that each carries the current in the same direction.

Helmholtz function (Helmholtz free energy) Symbol: F A thermodynamic function defined by $F = U - TS$, where U is the internal energy, T the thermodynamic temperature, and S the entropy. It is a measure of the ability of a system to do useful work in an isothermal process. *See also* free energy.

Helmholtz resonators A series of vessels of different sizes, each designed to resonate at only one particular frequency. They are used for detecting the overtones in musical notes.

henry Symbol: H The SI unit of inductance, equal to the inductance of a closed circuit that has a magnetic flux of one weber per ampere of current in the circuit. 1 H = 1 Wb A^{-1}.

hertz Symbol: Hz The SI unit of frequency, defined as one cycle per second (s^{-1}). Note that the hertz is used for regularly repeated processes, such as vibration or wave motion. An irregular process, such as radioactive decay, would have units expressed as s^{-1} (per second).

Hertzian waves An obsolete name for radio waves.

HF *See* high frequency.

high frequency (HF) A radio frequency in the range between 30 MHz and 3 MHz (wavelength 10 m–100 m).

high-temperature gas-cooled reactor *See* gas-cooled reactor.

high tension (H.T.) High voltage.

hole *See* semiconductor.

holography A method of recording (usually photographically) a three-dimensional image of an object. Normally laser light is used, but other radiations (including sound) can give holograms. The object is illuminated with laser light and the reflected light from the object is combined with direct light from the source, to give an interference pattern on a photographic plate. A three-dimensional image of the object is reconstructed by illuminating the interference pattern with the original light. The pattern includes information about phase and direction as well as intensity and colour.

homopolar generator (Faraday disc) A direct-current generator. It consists of a metal disc rotating in a magnetic field at right angles to the plane of the disc. The e.m.f. is induced between the centre of the disc and the edge.

Hooke's law The principle that if a body is deformed the strain produced is directly proportional to the applied stress. A graph of stress against strain for a material obeying Hooke's law is a straight line. When the stress is removed, the material returns to its original dimensions. Above a certain stress the material ceases to obey Hooke's law and the graph becomes non-linear. The point at which this occurs is the *proportional limit* of the material. *See also* elasticity.

horizontal intensity Symbol: B_H The strength of the Earth's magnetic field in a horizontal direction at a given point on or near the surface. In Britain the value of B_H is around 1.8×10^{-5} tesla; as at all other places the value is changing with time.
See also Earth's magnetism, magnetic variation.

horsepower Symbol: hp A unit of power equivalent to 745.700 watts.

hot-wire instrument An electrical measuring instrument in which the current is passed through a fine resistance wire, causing a rise in temperature. This may be measured by a thermocouple, or by the 'sag' in the wire caused by expansion. Hot-wire instruments can be used for alternating of currents without a rectifier. The deflection is roughly proportional to the square of the current.

H.T. High tension (voltage).

HTGR High-temperature gas-cooled reactor. *See* gas-cooled reactor.

hue A pure colour in the visible spectrum, relating to one visible wavelength. Any hue can be mixed with a certain other hue to produce the sensation of white. If a hue is mixed with white, the result is a tint. *See also* complementary colours.

humidity The amount of water vapour in the air. *Absolute humidity*, that is, the mass of water vapour per unit volume of air, is one way of expressing humidity. However, condensation and evaporation also depend on the air temperature, and so the *relative humidity* at a given temperature is more often used. Another measure is *specific humidity*, the mass of water vapour per unit mass of air. *See also* absolute humidity, relative humidity.

hundredweight An Imperial unit of mass equal to 112 lbs, (50.802 kg). Sometimes it is called the *long hundredweight* to distinguish it from the U.S. *short hundredweight* (100 lbs).

Huygens' construction A method of constructing a subsequent wavefront from an existing one. Each point on the original wavefront is thought of as a secondary point source, emitting radiation of the same frequency as the source, with a speed depending on the medium. The new wavefront is the surface tangential to all the *secondary wavelets* produced by those point surfaces in the forward direction. The construction can

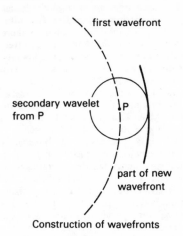

first wavefront

secondary wavelet
from P

P

part of new
wavefront

Construction of wavefronts

easily be used to show the effects of
rectilinear propagation, reflection, and
refraction. Huygens' principle also gives
good explanations for interference and
diffraction.

hybrid IC *See* integrated circuit.

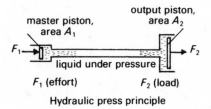

master piston,
area A_1

output piston,
area A_2

F_1

F_2

liquid under pressure

F_1 (effort) F_2 (load)

Hydraulic press principle

hydraulic press A machine in which
forces are transferred by way of pressure
in a fluid. In this case (and in the related
hydraulic braking system and hydraulic
jack) the force the user exerts (effort) is
less than the force the machine exerts
(load): the force ratio is greater than
one. The force ratio F_2/F_1 is equal to
A_2/A_1. This machine is not very efficient; frictional effects are large.

hydrogen electrode An electrode based

on hydrogen, and assigned zero electrode
potential, so that other elements may be
compared with it. The hydrogen is
bubbled over a platinum electrode,
coated in 'platinum black', in 1M acid
solution. Hydrogen is adsorbed on the
platinum black, enabling the equilibrium, $H(g) = H^+(aq) + e^-$, to be set
up. *See also* electromotive series.

hydrometer An instrument for measuring
the relative density of a fluid, usually a
liquid. The most common type is a
weighted glass bulb that, when floated
in a liquid, shows the liquid's relative
density by the depth of immersion. There
is usually a calibrated scale marked on
the glass.

hydrostatics The study of fluids (liquids
and gases) in equilibrium.

hygrometer An instrument for measuring
humidity. In a *chemical hygrometer*,
air is drawn through a drying agent.
Saturated air at the same temperature is
then drawn through an identical sample
of drying agent. The ratio of the different
amounts of water absorbed indicates
relative humidity. A *wet and dry bulb
hygrometer* consists of two ordinary
thermometers, one used normally and
the other with a bulb kept moist by a
damp wick. The cooling of the wet bulb
by evaporation depends on the humidity
of the air surrounding it.

hygroscope A device that indicates the
humidity of the air, often in the form of
a substance that changes colour in the
presence of moisture. *See also* hygrometer.

hyperfine structure *See* fine structure.

hypermetropia *See* hyperopia.

hyperons A group of elementary particles classified as baryons. They are
heavier than the nucleons (proton and
neutron) and have very short lifetimes.
See also elementary particles.

hyperopia (hypermetropia) Long sight

caused if the distance between the lens and the back of the eyeball is too short. Rays from a distant object would focus behind the retina if the eye were fully relaxed. The eye must accommodate to allow clear vision of distant objects. This means that light from close objects cannot be focused onto the retina. Long-sighted people have a distant near point. The defect is corrected with converging spectacle lenses.

hypsometer An instrument used for calibrating thermometers at the steam temperature. The thermometer bulb is placed above pure water boiling at a known pressure, and therefore a known temperature. *See* temperature scale.

hysteresis In general, an apparent lag of an effect behind whatever is causing it. *Magnetic hysteresis* is the behaviour of ferromagnetic materials as they are magnetized and demagnetized. The flux density, *B*, lags behind the external field strength *H*. *See* hysteresis cycle.

hysteresis cycle A closed loop obtained by plotting the flux density, *B*, of a ferromagnetic substance against the magnetizing field strength, *H*. The

substance is first brought to magnetic saturation from an unmagnetized state — this produces curve OA (see illustration). As the field strength is taken through one cycle of reductions, reversals, and increases, the curve follows the path ACDEFGA. This is known as a hysteresis loop.

The area of the loop equals the energy loss in taking the sample once through the cycle. This is known as *hysteresis loss* and depends on the substance.

OC (or OF) represents the *remanence*, the magnetic flux density remaining in the specimen when the saturating field is removed. OD (or OG) represents the *coercivity*, the value of the magnetizing field strength needed to reduce this remaining flux density to zero. All these details are readily explained by the domain behaviour of ferromagnetics.

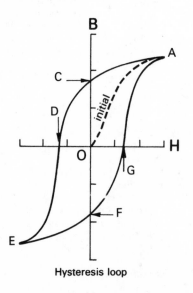

Hysteresis loop

I

IC *See* integrated circuit.

Iceland spar *See* calcite.

ice point *See* International Practical Temperature Scale.

ideal gas *See* gas laws, kinetic theory.

illumination Symbol: *E* A measure of the visible-radiation energy reaching a surface in unit time. Once called 'intensity of illumination', it is measured in lux (lx). One lux is an illumination of one lumen per square metre.
See also photometry.

image The point or region from which rays from an object appear to have come after reflection or refraction. Rays actually pass through a *real image* (like that produced on a screen by the lens of a view camera or slide projector). Rays do not actually pass through a *virtual image* (like that produced behind a make-up mirror or by a magnifying lens).

Image position is measured in terms of image distance (*v*) — the distance between the image and the lens or mirror. It relates to object distance (*u*) and focal distance (*f*) by the formula:
$1/f = 1/v + 1/u$ (real is positive, virtual is negative convention).
Truly sharp images are formed only with rays of a single wavelength passing fairly close and parallel to the principal axis of the lens or mirror.

To describe an image, three other factors are used in addition to its distance from the lens or mirror.
(1) Is it real or virtual?
(2) Is it upright or inverted — the same way up as the object or the other way up?
(3) How large is it compared to the object — magnified, same size, or diminished?
See also lens, magnification, mirror.

image plane The plane perpendicular to the axis centred on the image.

immersion objective A type of high-power microscope objective whose effectiveness is obtained by filling the space between lens and microscope slide with a suitable liquid. Traditionally oil is used, hence the common name *oil-immersion objective*. However, sugar solution is preferable. The liquid increases the effective aperture of the objective because of its high refractive constant. In turn this maximizes the resolution obtainable.

impedance Symbol: *Z* A measure of the opposition of a circuit to an alternating current. Impedance is the resultant of total reactance *X* and resistance *R*:
$$Z^2 = R^2 + X^2.$$
See also alternating current circuit, phasor, reactance.

Imperial units The system of measurement based on the yard and the pound, formerly used in the UK. The f.p.s. system was a scientific system based on Imperial units.

impulse (impulsive force) A force acting for a very short time, as in a collision. If the force is constant the impulse is $F\delta t$; if it is a variable force the impulse is the integral of this over the short time period. An impulse is equal to the change of momentum that it produces.

impulse noise *See* noise.

impulsive force *See* impulse.

incandescence The radiation of visible light by a surface at high temperature. Radiant electric fires and the filament lamp are domestic examples of incandescent sources. *See also* black body.

incidence, angle of *See* angle of incidence.

inclination (angle of dip) Symbol: δ The angle between the Earth's magnetic field and the horizontal at a given point on the Earth's surface. It is measured with an inclinometer and changes to some extent with time. *See also* Earth's magnetism, magnetic variation.

inclined plane A type of machine. Effectively a plane at an angle, it can be used to raise a weight vertically by movement up an incline.
Both distance ratio and force ratio depend on the angle of inclination. Efficiency can be fairly high if friction is kept low. The screw and the wedge are both examples of inclined planes.

inclinometer (dip circle) An instrument used to measure inclination (the angle of dip). A magnetic needle is pivoted so that it is free to move in a vertical plane in front of a circular scale. †The instrument is levelled and then rotated through a horizontal axis until the needle is vertical. The needle is now under the influence only of the vertical component of the Earth's field. When the dip circle is turned through 90° (into the meridian) the needle lines up along the Earth's magnetic field. Then the angle between the needle and the horizontal is the inclination at that point.

induced current Electric current in a conductor caused by e.m.f. set up by a changing magnetic field surrounding it. *See* electromagnetic induction.

inductance A measure of the e.m.f. produced in a circuit as a result of the magnetic effect of a changing electric current, either in that circuit or in another. *See* mutual induction, self-induction.

induction In general, the production of an effect by interaction with a field:
1. The production of magnetic order in a material by an external field. *See* magnetic induction.

2. The separation of electric charge in a material by an external electric field. *See* electrostatic induction.
3. The production of an e.m.f. (or resulting current) in a conductor in a changing magnetic field. *See* electromagnetic induction. *See also* mutual induction, self-induction.
4. *See* magnetic flux density.

induction coil A device that produces a series of high-voltage pulses by means of electromagnetic induction. It consists of a coil of wire with only a few turns, wound on an iron core and surrounded by another with many more turns. When the current in the first coil is interrupted suddenly, a large e.m.f. is induced in the second. A pulsed current in the first coil induces a large pulsed e.m.f. in the second.

induction heating (eddy-current heating) The temperature rise that occurs in conductors in a varying magnetic field. It is caused by the induced eddy-currents inside the block of material. Induction heating can be used for heating metals in a furnace. Some cookers work by heating the metal pots in this way. However, induction heating results in power losses from electrical machines, although usually this can be minimized with careful design.

induction motor An alternating-current electric motor in which the changing magnetic field of one coil, connected to the power supply, induces a current in another winding not connected to the supply. The magnetic force between the two coils turns the motor. Unlike most other electric motors, no brushes are needed to connect moving and stationary electrical parts. This eliminates sparking and therefore large motors are often of this type. *See also* electric motor, synchronous motor.

inelastic collision A collision for which the restitution coefficient is less than one. In effect, the relative velocity after the collision is less than that before; kinetic energy is not conserved in the collision, even though the system may be closed. Some of the kinetic energy is converted into internal energy. *See also* restitution, coefficient of.

inertia An inherent property of matter implied by Newton's first law of motion: 'a body continues in a state of rest or constant velocity unless acted on by an external force'. An object automatically opposes a change of motion by reason of its inertia. *See also* inertial mass, Newton's laws of motion.

inertial mass The mass of an object as measured by the property of inertia. It is equal to the ratio force/acceleration when the object is accelerated by a constant force. In a uniform gravitational field, it is equal to gravitational mass, since all objects have the same gravitational acceleration at the same place. *See also* gravitational mass.

inertial system A frame of reference in which an observer sees an object that is free of all external forces to be moving at constant velocity. The observer is called an *inertial observer*. Any frame that moves with constant velocity and without rotation relative to an inertial frame is also an inertial frame. Newton's laws of motion are valid in any inertial frame (but not in an accelerated frame), and the laws are therefore independent of the velocity of an inertial observer. *See also* frame of reference, Newton's laws of motion.

infrared (IR) Electromagnetic radiation past the red end of the visible spectrum. The range of wavelengths is approximately 0.7 μm to 1 mm. Infrared is sometimes called *thermal radiation*; most of the energy radiated by surfaces below about 6000 K is the infrared. Many materials transparent to visible light are opaque to infrared, including glass. Rock salt, quartz, germanium, or polyethylene prisms and lenses are suitable for use with infrared. Infrared radiation close to visible light in wavelength can be detected by thermometers with blackened bulbs and with photographic film. Otherwise a thermocouple or a bolometer can be used.

Infrared radiation is produced by movement of charges on the molecular scale; i.e. by vibrational or rotational motion of molecules. *See also* electromagnetic spectrum.

infrasound Vibrations in a medium with frequencies below about 16 hertz. The ear distinguishes such vibrations as a series of pulses, rather than a continuous sound.

instantaneous value The value of a varying quantity (e.g. acceleration, current, power, etc.) at a particular instant in time.

insulation, electrical The use of nonconductors to coat or separate electrical conductors (for safety or prevention of short circuits). Rubber, PVC, and ceramics are common insulators.

insulation, thermal Any means of preventing or reducing the transfer of thermal energy. Lagging of hot-water tanks and foam-filled cavity walls are examples. *See also* insulator, thermal.

insulator A material with a very high electrical resistivity. Most non-metallic materials are good electrical insulators. *Compare* conductor (thermal), semiconductor.

insulator, thermal A material that does not readily transmit heat. Many insulators, such as asbestos, are porous or fibrous solids that trap small pockets of air within them. Gases do not conduct heat well, and convection is prevented because the air cannot flow. *See also* conductor (thermal).

integrated circuit (IC) A circuit that incorporates numerous components into one unit. In a *monolithic IC* a single chip of silicon is the base or substrate onto which all the individual components are integrated during manufacture. *Hybrid ICs* consist of one or more monolithic ICs mounted on a substrate, or several components similarly mounted and interconnected. After manufacture neither type can be dismantled. decibel.

intensity A measure of the rate of energy transfer by radiation. The unit of intensity is the watt per square metre (W m^{-2}).
1. The intensity of visible radiation is related to brightness. In *photometry* special units are used because it is necessary to consider the sensitivity of the eye to different visible radiations. Thus the intensity of a light source — sometimes called luminous intensity — is measured with the candela. The intensity of illumination of a surface is measured in lux. The intensity of visible radiation itself is measured in lumens.
2. The intensity of sound relates to the sensation of loudness though this depends also on frequency. It is inversely proportional to the square of the distance from the source. The *intensity level* of a sound is the intensity relative to an agreed standard. *See* decibel.

intensity, electric *See* electric field strength.

intensity, magnetic *See* magnetic field strength.

interaction A mutual effect between two or more systems or bodies, so that the overall result is not simply the sum of the separate effects. There are four separate interactions distinguished in physics:
(1) *Gravitational interaction* The weakest of the four, about 10^{40} times weaker than the electromagnetic interaction. It is an interaction between bodies or particles on account of their mass, and operates over long distances. *See* Newton's law of universal gravitation.
(2) *Electromagnetic interaction* The interaction between charged bodies or particles (stationary or moving). It falls off with the square of distance and operates over all distances. *See also* Coulomb's law.

(3) *Strong interaction* An interaction between hadrons, about 100 times greater than the electromagnetic interaction. It operates at very short range (up to around 10^{-15} m) and is the force responsible for holding nucleons together in the atomic nucleus.
(4) *Weak interaction* An interaction about 10^{10} times weaker than the electromagnetic interaction. It occurs between leptons, and is the interaction in beta decay.
So far it has not proved possible to formulate a unified field theory for all four types of interaction, although there has been some success in unifying the electromagnetic and weak interactions.

interference The effect of similar waves passing through the same region. At each point the amplitude is the sum of the amplitudes of each (taking signs into consideration), perhaps producing a static pattern. The waves emerge from the overlapping region unaffected. A static interference pattern consists of a network of points of *constructive interference* (where the waves reinforce each other), and of points of *destructive interference* (where they cancel each other). Interference effects are observed with all types of waves. When first observed with light (Young, 1801), the wave theory of light appeared proved.
Interference patterns are obtained only with coherent radiation: the two wave trains have similar amplitudes and a constant phase relationship (and, therefore, equal wavelengths). Transverse waves must both be unpolarized or polarized in the same plane. There are two basic methods of producing interference patterns. Division of wavefront involves producing two coherent sources from a single source by a method such as Young's double slit, Fresnel's biprism, or Lloyd's mirror. Division of amplitude methods involve reflecting part of a wave and transmitting part, as in Newton's rings, the air wedge, and thin films.

interferometry The technique of producing interference patterns and

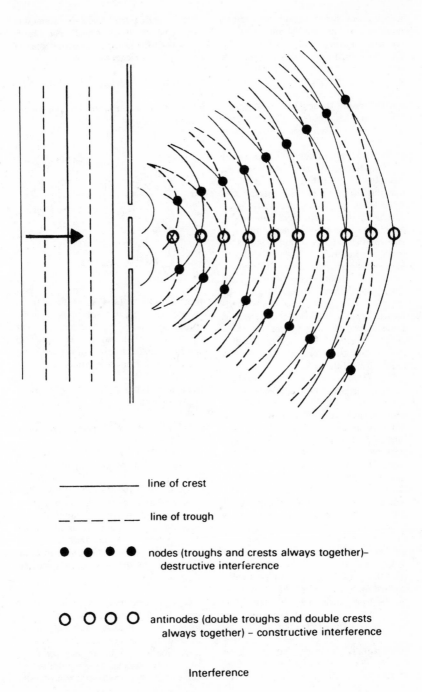

line of crest

line of trough

nodes (troughs and crests always together)–
destructive interference

antinodes (double troughs and double crests
always together) – constructive interference

Interference

using these for accurate wavelength measurement, measuring small distance changes, testing flat surfaces, etc. Devices, such as etalons, used for producing and using interference patterns are *interferometers*.

intermolecular forces Forces of attraction between molecules or neutral atoms. *See* van der Waals forces.

internal energy Symbol: U The energy of a system that is the total of the kinetic and potential energies of its constituent particles (e.g. atoms and molecules). If the temperature of a substance is raised, by transferring energy to it, the internal energy increases (the particles move faster). Similarly, work done on or by a system results in an increase or decrease in the internal energy. The relationship between heat, work, and internal energy is given by the first law of thermodynamics. Sometimes the internal energy of a system is loosely spoken of as 'heat' or 'heat energy'. Strictly, this is incorrect; heat is the transfer of energy as a result of a temperature difference.

internal resistance Resistance of a source of electricity. In the case of a cell, when a current is supplied, the potential difference between the terminals is lower than the e.m.f. The difference (i.e. e.m.f. - p.d.) is proportional to the current supplied. The internal resistance (r) is given by:
$$r = (E - V)/I$$
where E is the e.m.f., V the potential difference between the terminals, and I the current.

international candle A former unit of luminous intensity, approximately equal to 1.018 3 candela. It was originally defined in terms of the light emitted per second by a specified electric lamp, but after 1946 was superseded by the candela, now the SI base unit.

International Practical Temperature Scale (IPTS) A scale of temperatures (1968) based on a number of fixed points, with agreed methods of measuring temperatures over particular ranges. It is used for practical measurements of thermodynamic temperature.

interstitial *See* defect.

intrinsic semiconductor *See* semiconductor.

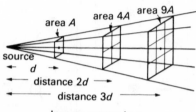

Inverse square law

inverse square law A relationship in which the effect of a source is inversely proportional to the square of the distance from the source. It applies to energy radiated from a point source and also to fields from point sources.

inversion temperature *See* Joule-Kelvin effect.

inverter gate (NOT gate) *See* logic gate.

inverting prism *See* erecting prism.

ion A charged particle consisting of an atom, or group of atoms, that has either lost or gained electrons. Sodium chloride (salt) for example, is made up of positive sodium ions — atoms that have each lost an electron — and negative chlorine ions — atoms that have each gained an electron. *See* ionization.

ionization The process of producing ions. There are several ways in which ions may be formed from atoms or molecules. In certain chemical reactions ionization occurs by transfer of electrons; for example, sodium atoms and chlorine atoms react to form sodium

International Practical Temperature Scale

Fixed point	TK	°C
triple point of hydrogen	13.81	−259.34
hydrogen with vapour pressure 25/76 atmosphere	17.042	−256.108
boiling temperature of hydrogen	20.28	−252.87
boiling temperature of neon	27.102	−246.048
triple point of oxygen	54.361	−218.789
boiling temperature of oxygen	90.188	−182.962
triple point of water	273.16	0.01
boiling temperature of water	373.15	100
melting temperature of zinc	692.73	419.58
melting temperature of silver	1235.08	961.93
melting temperature of gold	1337.58	1064.43

Note: the scale given above is based on standard interpolation procedures to be used experimentally for measurements between the fixed points. The methods employed are:

−259.35°C to 630.74°C   resistance measurement with a platinum resistance thermometer

630.74°C to 1064.43°C   e.m.f. measurement with a thermocouple made of platinum and platinum/10% rhodium alloy

above 1064.43°C   radiation measurement and use of Planck radiation law

chloride, which consists of sodium ions (Na^+) and chloride ions (Cl^-). Certain molecules can ionize in solution; acids, for example, form hydrogen ions as in the reaction

$$H_2SO_4 \rightarrow 2H^+ + SO_4^{2-}$$

Ions can also be produced by ionizing radiation; i.e. by the impact of particles or photons with sufficient energy to break up molecules or detach electrons from atoms: $A \rightarrow A^+ + e^-$. Negative ions can be formed by capture of electrons by atoms or molecules: $A + e^- \rightarrow A^-$.

ionization chamber A chamber in which ionizing radiation can be detected and/or measured. It contains a pair of electrodes between which is maintained a potential difference. Passage of ionizing radiation through the chamber produces electrons and positive ions, which separate and move to the electrode of opposite sign. There is thus a small electrical current between the electrodes, which relates to the intensity of the radiation.

ionization potential Symbol: I The energy required to remove an electron from an atom or molecule to form a positive ion.

ionizing radiation Radiation of sufficient energy to cause ionization of substances through which it passes. Ionizing radiations include high-energy charged particles and electromagnetic radiation such as gamma radiation, X-radiation, or short-wavelength ultraviolet radiation.

ionosphere A region in the Earth's upper atmosphere containing ions and free electrons. It is formed by ionization by ultraviolet radiation from the Sun. It is divided into three layers: D-layer (50-90 km), E-layer (90-150 km), and F-layer (150-1000 km). Radio transmission can occur by reflection from the ionosphere.

ion pair A pair of ions of opposite charge produced, for example, by an ionizing radiation: $AB \rightarrow A^+ + B^-$.

IPTS *See* International Practical Temperature Scale.

IR *See* infrared.

iris An arrangement able to vary the amount of light that enters an optical instrument. In the mammalian eye, this is a circular muscle that changes the size of the pupil. In many optical instruments, a similar effect is obtained with a diaphragm. In either case the aperture is varied.

iris diaphragm *See* diaphragm.

irradiance Symbol: E The rate of energy reaching unit area of a surface; i.e. the radiant flux per unit area. Unlike illumination, irradiance is not restricted in use to visible radiation. The unit is the watt per square metre (W m^{-2}).
The *solar constant* is the irradiance of the Earth by the Sun. The value is 1.35 KW m^{-2}. When applying this figure to a consideration of solar power, allowance must be made for atmospheric absorption and reflection, and for the angle of the Sun from the zenith.

irreversible change *See* reversible change.

isobars 1. Two or more nuclides that have the same nucleon numbers but different proton numbers.
2. Lines joining points of equal pressure.

isoclinic line An imaginary line on the Earth's surface joining points with the same inclination (angle of dip). The *magnetic equator* (or *aclinic line*) is the line that joins points of zero inclination. *See also* Earth's magnetism.

isodynamic line An imaginary line on the Earth's surface joining points at which the Earth's total magnetic field strength is the same. *See also* Earth's magnetism.

isogonal See isogonic line.

isogonic line (isogonal) An imaginary line on the Earth's surface joining points of equal declination. Isogonic lines run roughly East to West. *Agonic lines* are isogonic lines that join points having zero declination. *See also* Earth's magnetism.

isolated system *See* closed system.

isomers, nuclear *See* nuclear isomers.

isotherm A line on a chart or graph joining points of equal temperature. *See also* isothermal change.

isothermal change A process that takes place at a constant temperature. The system is in thermal equilibrium with its surroundings throughout an isothermal process. For example, a cylinder of gas in contact with a constant-temperature box may be compressed slowly by lowering a piston. The work done appears as energy, which flows into the reservoir to keep the gas at the same temperature. Isothermal changes are contrasted with *adiabatic* changes, in which no energy enters or leaves the system, and the temperature changes. In practice no process is perfectly isothermal and none is perfectly adiabatic. Many can approximate in behaviour to one of these ideals.

isotones Two or more nuclides that have the same neutron numbers but different proton numbers.

isotopes Two or more species of the same element differing in their mass numbers because of differing numbers of neutrons in their nuclei. The nuclei must have the same number of protons (an element is characterized by its proton number). Isotopes of the same element have very similar chemical properties (the same electron configuration), but differ slightly in their physical properties. An unstable isotope is termed a *radioactive isotope* or *radioisotope*. For example, potassium has 3 naturally occurring isotopes of mass numbers 39, 40, and 41 respectively.
$^{39}_{19}$K has 19 protons and 20 neutrons

${}^{40}_{19}$K has 19 protons and 21 neutrons
${}^{41}_{19}$K has 19 protons and 22 neutrons
${}^{40}_{19}$K is radioactive with a half-life of 1.3 $\times 10^9$ years.
At least eight other isotopes of potassium can be produced by nuclear reactions but all are highly unstable (radioactive).

isotope separation The process of separating the different isotopes of an element according to their atomic masses. Most methods depend on small differences in physical properties between isotopes.

On the laboratory scale isotopes of elements can be separated in the mass spectrometer in which the paths of nuclei depend on the ratio of charge to mass. Similar methods have been used on a larger scale. The difference in rates of diffusion of the volatile compounds ^{238}uranium hexafluoride and ^{235}uranium hexafluoride has been used on a large scale to separate the isotopes of uranium. The ultracentrifuge process is also used. ${}^{235}UF_6$ tends to stay near the axis of the centrifuge and the ${}^{238}UF_6$ tends to drift to the perimeter. Slight separation occurs at each stage.
Early methods of isolating deuterium used electrolysis of water. Deuterium ions are discharged at a slightly slower rate; thus the residue becomes progressively richer in deuterium during electrolysis. The process presently used is a chemical exchange between water and hydrogen sulphide.

isotopic mass (isotopic weight) The mass number of a given isotope of an element.

isotopic number The difference between the number of neutrons in an atom and the number of protons.

isotopic weight *See* isotopic mass.

isotropy A medium is isotropic if the value of a measured physical quantity does not depend on the direction. *Compare* anisotropy.

Joly steam calorimeter An apparatus for measuring the specific thermal capacity of a gas at constant volume. Two identical copper globes are suspended from the arms of a balance — one is evacuated and the other filled with the sample. The two globes are surrounded by a container into which steam is passed. More condensation occurs on the filled globe because of the higher thermal capacity. The specific heat capacity (C_v) of the gas can be calculated from
$$mC_v(\theta_2 - \theta_1) = m_s L$$
where m is the mass of gas, θ_2 the temperature of the steam, θ_1 the initial temperature of the globes, m_s the mass of steam condensed, and L the specific latent thermal capacity of evaporation of water.

joule Symbol: J The SI unit of energy and work, equal to the work done when the point of application of a force of one newton moves one metre in the direction of action of the force. 1 J = 1 N m. The joule is the unit of all forms of energy.

Joule–Kelvin effect (Joule–Thomson effect) A temperature change that occurs when a gas expands through a porous barrier into a region of lower pressure. Most real gases, when expanded in this way, are cooled slightly because they do work against their own intermolecular forces.

An ideal gas would not show the Joule–Kelvin effect because there are no intermolecular forces. The phenomenon is the basis of a method of liquefying gases. The temperature fall depends on the difference in pressure across the plug.
The cooling shows a deviation from Joule's law. In addition, deviations of the gas behaviour from Boyle's law may cause either increase or decrease of temperature. At a certain temperature

Joule's equivalent

— the *inversion temperature* of the gas
— the temperature rise from the second effect equals the drop produced by deviation from Joule's law, and there is no change in temperature. Below its inversion temperature a gas is cooled by the expansion; above it it is heated.

Joule's equivalent Symbol: J A constant, 4.1855×10^7 ergs per calorie (15°), relating former units of 'heat' energy to units of mechanical energy. It arose from early experiments by Joule showing that mechanical work always relates to an equivalent quantity of 'heat'. The constant is also called the *mechanical equivalent of heat*.
In SI units, work, heat, and all forms of energy are measured in joules. The value of J (4.1855 joules per calorie) is the factor defining the calorie.

Joule's laws 1. Relationships between the 'heating effect' of an electric current in a conductor to the electrical quantities concerned. Joule found that the energy transferred per unit time (the power) in a given conductor was proportional to the square of the current. He also showed that it was proportional to resistance, for a given current. The results can be combined in the relationship

$$P = I^2R$$

where P is the power (in watts), I the current (in amperes), and R the resistance (in ohms). See also Joule's laws.
2. The internal energy of an ideal gas is independent of its volume under adiabatic conditions. For real gases this is not strictly true because the internal energy depends on the potential energies of the particles of gas. Thus, intermolecular forces cause changes in internal energy when a volume change occurs. See also Joule–Kelvin effect.

Joule–Thomson effect See Joule–Kelvin effect.

junction rectifier See rectifier.

junction transistor See transistor.

K

kaon (K-meson) A type of meson. There are two types of kaon, having negative or positive charge. *See also* elementary particles.

keepers Pieces of 'soft' iron placed between the poles of permanent magnets when stored to help retain the magnetism. Bar magnets are placed in pairs with opposite poles next to each other and keepers are placed across each end. The magnet's internal field opposes its external field; this causes a demagnetizing effect. When keepers are used, they become magnetized with an internal field in the same direction as that of the magnet. Thus the demagnetizing effect is reduced. *See also* magnetic induction.

kelvin Symbol: K The SI base unit of thermodynamic temperature. It is defined as the fraction 1/273.16 of the thermodynamic temperature of the triple point of water. Zero kelvin (0 K) is absolute zero. One kelvin is the same as one degree on the Celsius scale of temperature.

Keplerian telescope The most common type of refracting telescope arrangement, consisting of converging objective and eyepiece. Unlike the Galilean telescope, it provides an inverted image and has a greater length. However the field of view is larger and image quality is higher.
The angular magnification is given by f_O/f_E; the lens separation is $(f_O + f_E)$. For terrestrial use an inverting lens can be included between the objective and the eyepiece. This is placed so that it is a distance $2f_I$ (f_I is its focal distance) from the first image formed by the objective. An erect image is formed a distance $2f$ behind the inverting lens, arranged to be at the principal focus of the eyepiece. The magnification is not

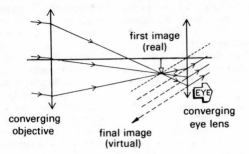

Keplerian telescope

affected but the distance from objective to eyepiece is now $f_O + f_E + 4f$.
See also refractor.

Kepler's laws The laws of planetary motion deduced in about 1610 by Johannes Kepler using astronomical observations made by Tycho Brahe:
(1) the planets describe elliptical orbits with the Sun at one focus of the ellipse.
(2) the line between a planet and the Sun sweeps out equal areas in equal times.
(3) the square of the period of a planet's orbit is proportional to the cube of the semi-major axis of the ellipse.
Application of the third law to the orbit of the Moon about the Earth gave support to Newton's theory of gravitation.

Kerr effect The appearance of birefringence in certain isotropic substances when placed in a strong electric field. The effect is proportional to the square of the field strength. It is used in *Kerr cells*, which can switch light radiation extremely rapidly (often within a few picoseconds). Benzene is an example of a substance showing the Kerr effect.

keV Kiloelectronvolts; 10^3 electronvolts.

kilo- Symbol: k A prefix denoting 10^3. For example, 1 kilometre (km) = 10^3 metres (m).

kilogram (kilogramme) Symbol: kg The SI base unit of mass, equal to the mass of the international prototype of the kilogram, which is a piece of platinum-iridium kept at Sèrres in France.

kilogramme An alternative spelling of *kilogram*.

kilowatt-hour Symbol: kwh A unit of energy, usually electrical, equal to the energy transferred by one kilowatt of power in one hour. It is the same as the Board of Trade unit and has a value of 3.6×10^6 joules.

kinematics The study of the motion of objects. *See also* dynamics.

kinematic viscosity Symbol: ν The ratio of a fluid's viscosity to its density. *See also* viscosity.

kinetic energy Symbol: T The work that an object can do because of its motion. For an object of mass m moving with velocity v, the kinetic energy is given by $mv^2/2$. This gives the work the object would do in coming to rest. The rotational kinetic energy of an object of moment of inertia I and angular velocity ω is given by $I\omega^2/2$.
See also energy.

kinetic friction *See* friction.

kinetic theory A theory explaining physical properties in terms of the motion of particles.

kinetic theory (of gases) The molecules or atoms of a gas are in continuous random motion and the pressure (p) exerted on the walls of a containing vessel arises from the bombardment by these fast moving particles. These have average speeds, at normal pressures and temperatures, of around one kilometre per second. When the temperature is raised these speeds increase; so consequently does the pressure. If more particles are introduced or the volume is reduced there are more particles to bombard unit area of the walls and the pressure also increases. When a particle collides with the wall it experiences a rate of change of momentum, which is equal to the force exerted. For a large number of particles this provides a steady force per unit area (or pressure) on the wall.

Following certain additional assumptions, the kinetic theory leads to an expression for the pressure exerted by an ideal gas. The assumptions are:

(1) The particles behave as if they are hard smooth perfectly elastic points.

(2) They do not exert any appreciable force on each other except during collisions.

(3) The volume occupied by the particles themselves is a negligible fraction of the volume of the gas.

(4) The duration of each collision is negligible compared with the time between collisions.

By considering the change in momentum on impact with the walls it can be shown that $p = \rho c^2 3$ where ρ is the density of the gas and c is the root-mean-square speed of the molecules. The mean square speed of the molecules is proportional to the absolute temperature:

$$Nmc^2 = RT$$

See also degrees of freedom, equipartition of energy.

Kirchhoff's laws A set of rules for calculating unknown currents, resistances, and voltages in an electric circuit. They are:

(1) The algebraic sum of the currents at any point in any circuit is zero. For example, if 6 amperes enter a three-way junction through one wire, then 6 amperes must leave through the other two. A current flowing away from a junction has an opposite sign to one flowing towards the junction.

(2) The algebraic sum of the e.m.f.s round any closed loop in any circuit is equal to the sum of the products of current and resistance around the loop. For example, in a circuit with e.m.f. E, current I, and resistances R and r, $E = Ir + IR$.

Kirchoff's law For a given wavelength the emissivity of a surface in a particular direction is equal to the absorptance for radiation incident from that direction.

K-meson *See* kaon.

knife edge A sharp wedge used as a fulcrum or support, as in a balance. The sharp tip minimizes the area of contact between moving parts, thereby reducing the friction between them. Knife edges are made of hard material such as agate.

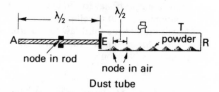

Dust tube

Kundt's tube A device for showing stationary waves in a gas (or liquid). A horizontal rod, clamped at its midpoint, has a flat disc on one end. The disc just fits into the bore of a glass tube, which is closed at the other end with a flat surface. Fine powder is sprinkled along the inside of the tube. The rod is stroked to produce longitudinal vibrations; sound waves are generated by the disc, travel down the tube, and are reflected at the closed end. The position of the disc is changed (changing the length of the column of gas) until standing waves are produced. The dust then vibrates strongly and settles in regularly-spaced heaps along the tube. These are *nodes*.

The device can be used to find the speeds of sound in different materials. It can be shown that:

$$V_r = l_r V_g / l$$

where V_r is the speed of sound in the rod, V_g the speed in the gas, l_r the length of the rod, and l the distance between nodes in the gas. Similar experiments can be performed by replacing the rod and disc with a loudspeaker connected to a variable signal generator.

L

label A stable or radioactive nuclide used to investigate some process, such as a chemical reaction. *See* radioisotope.

laevorotatory *See* optical activity.

lag The time or angle by which one periodic quantity is delayed with respect to another.

lambda particle *See* elementary particles.

lambert A former unit of luminance. It is approximately 3.18×10^3 Cd m^{-2}.

Lambert's law 1. First proposed in 1760; then restricted to visible light, it is now used with all radiations. The law concerns the rate of absorption of radiation as it travels deeper into a medium. It states that equal thicknesses of the medium absorb equal proportions of the incident radiation. In other words, the intensity I of the transmitted radiation falls off exponentially with distance d in the medium:

$$I = I_0 \exp{-\alpha d}$$

Here I_0 is the intensity of the initially incident radiation, and α is the *linear absorption coefficient* of the medium.

As well as depending on the medium, α varies with wavelength.

2. In photometry, the fact that the luminous intensity of a diffuse surface varies with angle of view:

$$I_0 = I_0 \cos\theta$$

Here, I_0 is the intensity along the normal, while I is that along a line at angle θ to the normal. The principle is often called *Lambert's cosine law*.

laminar flow Steady flow in which the fluid moves past a surface in parallel layers of different velocities. *Compare* turbulent flow. *See also* Poiseuille's formula.

laminated iron A piece of iron constructed in thin layers that are separated by electrical insulator. Laminated iron cores are used in many electric machines. The laminations reduce eddy currents caused by a changing magnetic field.

laser (*l*ight *a*mplification by *s*timulated *e*mission of *r*adiation) A device able to produce a beam of radiation with unusual properties. Generally the beam is: coherent (the waves are in phase); monochromatic (the waves are of effectively the same wavelength); parallel; and intense (carrying a great deal of energy).

There are innumerable applications of such beams in communications, engineering, science, and medicine. Laser action is obtained in a volume of suitable material (solid, liquid, or gas) into which energy is passed at a high rate.

The input energy excites the active particles to a higher energy state W_2 from which they return to a comparatively stable state W_1 above the ground state, W_0. They accumulate there, forming a *population inversion*. Passing photons of energy (W_1-W_0) stimulate decay to the ground state. The photons emitted travel in phase with, and in the same direction as, those that stimulated their production.

In practice, reflecting surfaces are used at each end of the device — one totally reflecting and the other partially reflecting. The radiation is reflected back-

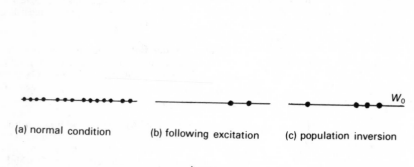

(a) normal condition   (b) following excitation   (c) population inversion

Laser

wards and forwards, building up an intense beam, which is emitted through the partially reflecting surface.

In solid lasers, such as the ruby laser, the population inversion is produced by an intense external light source. Generally it is pulsed. The wavelength is 694.3 nm. In gas lasers, a discharge is used. The carbon dioxide laser gives a wavelength of 10.5 μm. Helium–neon lasers produce a number of separate wavelengths, including 1.153 μm, 3.391 μm, and 632.8 μm. Gas lasers can produce continuous (rather than pulsed) beams.

See also holography.

latent heat The energy transferred when a substance changes its state from solid to liquid, liquid to gas, solid to gas, etc. For example, when a liquid boils the latent heat taken in is partly used to overcome attractive forces between the molecules, and partly to do work in expanding against atmospheric pressure. The *specific latent heat* (specific latent thermal capacity) is the energy transferred in converting unit mass from one state to another at the same temperature. For instance, the specific latent thermal capacity of fusion (solid-liquid) or the specific latent thermal capacity of evaporation (liquid-gas). The units are joules per kilogram (J kg^{-1}). Latent heats are standard enthalpies for these processes.

lateral inversion *See* perversion.

lattice *See* crystal.

law of flotation *See* flotation, law of.

law of magnetic poles (law of magnetism) The rule describing the forces between nearby poles — like poles repel each other; unlike poles attract each other.

law of moments *See* moment.

laws of conservation *See* conservation of mass and energy, constant energy (law of), constant (linear) momentum (law of), constant mass (law of), constant angular momentum (law of).

laws of electromagnetic induction *See* electromagnetic induction.

laws of friction *See* friction.

laws of reflection *See* reflection, laws of.

laws of refraction *See* refraction, laws of.

laws of thermodynamics *See* thermodynamics.

LCR circuit *See* alternating-current circuit.

lead-acid accumulator A type of electrical accumulator used in vehicle batteries. It has two sets of plates: spongy lead plates connected in series to the negative terminal and lead oxide plates connected to the positive terminal. The material of the electrodes is held in a hard lead-alloy grid. The plates are interleafed. The electrolyte is dilute sulphuric acid. The e.m.f. when fully charged is about 2.2 V. This falls to a steady 2 V when current is drawn. As the accumulator begins to run down, the e.m.f. falls further. During discharge the electrolyte becomes more dilute and its relative density falls. To recharge the accumulator, charge is passed through it in the opposite direction to the direction of current supply. This reverses the cell reactions, and increases the relative density of the electrolyte (c. 1.25 for a fully charged accumulator).

The electrolyte contains hydrogen ions (H^+) and sulphate ions (SO_4^{2-}). During discharge, H^+ ions react with the lead(IV) oxide to give lead(II) oxide and water

$$PbO_2 + 2H^+ + 2e^- \rightarrow PbO + H_2O$$

This reaction takes electrons from the plate, causing the positive charge. There is a further reaction to yield soft lead sulphate:

$$PbO + SO_4^{2-} + 2H^+ \rightarrow PbSO_4 + H_2O + 2e^-$$

Electrons are released to the electrode, producing the negative charge.

During charging the reactions are reversed:

$$PbSO_4 + 2e^- \rightarrow Pb + SO_4^{2-}$$
$$PbSO_4 + 2H_2O \rightarrow PbO_2 + 4H^+ + SO_4^{2-} + 2e^-$$

Leclanché cell A primary voltaic cell consisting, in its 'wet' form, of a carbon-rod anode and a zinc cathode, with a 10–20% solution of ammonium chloride as electrolyte. Manganese dioxide mixed with crushed carbon in a porous bag or pot surrounding the anode acts as a depolarizing agent. The dry form (*dry cell*) is widely used for torch batteries, transistor radios, etc. It has a mixture of ammonium chloride, zinc chloride, flour, and gum forming an electrolyte paste. Sometimes the dry cell is arranged in layers to form a rectangular battery, which has a longer life than the cylinder type.

LED *See* light-emitting diode.

Lees' disc An apparatus for measuring the thermal conductivity of a poor conductor. The sample is a thin flat disc of known area (A) and thickness (l) placed between two discs of brass (or other good conductor), each having a thermometer in a hole in its side. The apparatus is suspended by strings and the top disc is raised in temperature. Under steady conditions energy passes at a steady rate and the temperatures of the top and bottom metal discs are θ_2 and θ_1. The bottom conductor loses energy by radiation: to measure the rate of loss of energy, the bottom slab is warmed to the temperature θ_2, the insulator placed on top, and a cooling curve plotted so that rate of temperature fall can be obtained at the lower temperature θ_1. Then k is found from the equation

$$kA(\theta_2 - \theta_1)/l = mc\,d\theta/dt$$

where m is the mass of the bottom slab, c its specific thermal capacity, and $d\theta/dt$ its rate of temperature fall obtained from the cooling curve.

The apparatus can be modified for use with liquids, these being contained in a narrow insulating ring held between the two metal discs.

left-hand rule *See* Fleming's rules.

lens An optical component that refracts rays passing through, and either converges or diverges them. With visible radiation, glass lenses in air are normally used; however lenses of any transparent substance set in any other would work. A lens is best described by its effect on a parallel beam. Converging lenses make it convergent; diverging lenses make it divergent. Focal distance is the usual measure of the power of a lens to converge a parallel beam. Parallel rays are refracted by converging and diverging lenses as shown in the diagram. Strictly, such focusing occurs only with rays parallel to and close to the principal axis. Rays from an object will be refracted so that they appear to come from somewhere else — the image. Two examples are given here to show how ray diagrams are drawn.

The object distance (u) and the image distance (v) relate to the focal distance (f) thus:
$1/f = 1/v + 1/u$ (real is positive, virtual is negative sign convention).

The formula applies to real or virtual focal points, objects, and images. All the above can be adapted to lenses for use with nonvisible radiations, including pressure waves and electron beams.

Lenz's law The direction of an induced e.m.f. is such that it opposes the change that produces it. For example, if a permanent bar magnet is moved towards a coil that forms part of a complete circuit, then current flows so that the magnetic field around the coil repels the magnet. If this were not the case, no work would need to be done to cause charge to flow, breaking the principle of constant energy.

leptons A group of elementary particles including the electron, the muon, and the neutrino. They are distinguished from hadrons by their type of interaction. *See also* elementary particles.

Leslie's cube A cube-shaped can with the four sides painted different colours or given different finishes (polished, rough, etc.). It is filled with hot water and used in experiments on the effect of surface on emissivity.

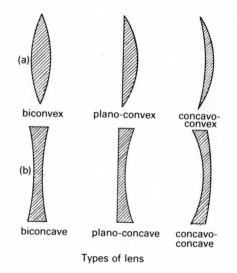

(a)

biconvex   plano-convex   concavo-convex

(b)

biconcave   plano-concave   concavo-concave

Types of lens

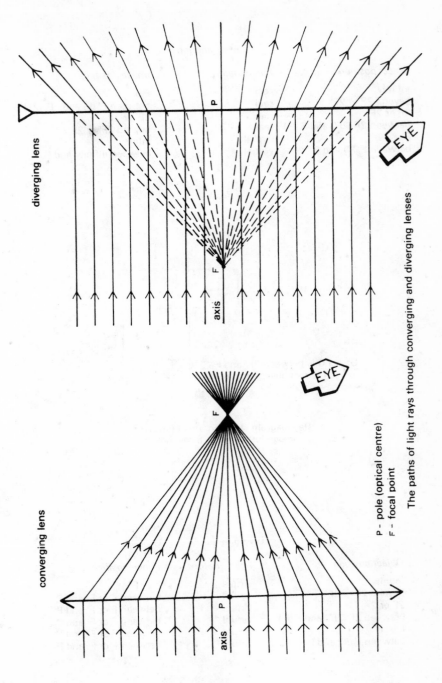

converging lens

diverging lens

axis

axis

P

F

P

F

EYE

EYE

P - pole (optical centre)
F - focal point

The paths of light rays through converging and diverging lenses

Converging lens

Object position	Image nature	Image position
infinity	real	F or F'
beyond 2F or 2F'	real	between F' (F) and 2F' (2F)
2F or 2F'	real	2F' or 2F
between 2F (2F') and F (F')	real	beyond 2F' or 2F
F or F'	virtual	infinity
between F (F') and P	virtual	between infinity and P

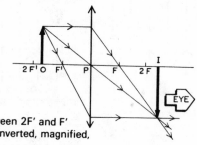

Object O is between 2F' and F'
Image I is real, inverted, magnified,
and beyond 2F

Ray diagram showing a real (inverted)
image formed from a real object

Diverging lens

Object position	Image nature	Image position
infinity	virtual	F or F'
beyond 2F or 2F'	virtual	between F or F' and P
2F or 2F'	virtual	between F or F' and P
between 2F (2F') and F (F')	virtual	between F or F' and P
F or F'	virtual	between F or F' and P
between F (F') and P	virtual	between F or F' and P

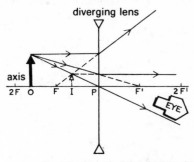

image I is between
F and P, virtual,
upright, and
diminished

Ray diagram showing a virtual (erect)
image formed from a real object

lever A class of machine. Levers are rigid objects able to turn around some point (pivot or fulcrum). The force ratio and the distance ratio depend on the relative positions of the pivot, the point where the user exerts the effort, and the point where the lever applies force to the load. There are three types (orders) of lever; each can be found single or double as shown.

Levers can have high efficiency; the main energy losses are by friction at the pivot, and by moving the lever itself.

The range of colours in the visible spectrum

Wavelength range (nm)	Colour
400 – 420	violet
420 – 450	indigo
450 – 500	blue
500 – 550	green
550 – 600	yellow
600 – 650	orange
650 – 760	red

Note that perception of different colours varies between individuals.

Leyden jar An early type of capacitor made from a glass jar with metal foil on its inner and outer surfaces.

LF *See* low frequency.

light (visible radiation) A form of electromagnetic radiation able to be detected by the human eye. Its wavelength range is between about 400 nm (far 'red') and about 700 nm (far 'violet'). The boundaries are not precise as individuals vary in their ability to detect extreme wavelengths; this ability also declines with age.

Light is produced by surfaces at temperatures above about 900 K. Below about 6000 K, however, the majority of the radiation emitted is infrared. The standard household filament lamp works at about 2800 K; at this temperature only a few percent of the emitted radiation is visible. 'Cold' sources, such as certain chemical reactions, glow worms, lasers, and discharge tubes are more efficient in this sense.

These produce light by specific transitions between electronic energy levels in atoms and molecules. Their spectra are therefore not continuous, but sets of lines.

light-emitting diode

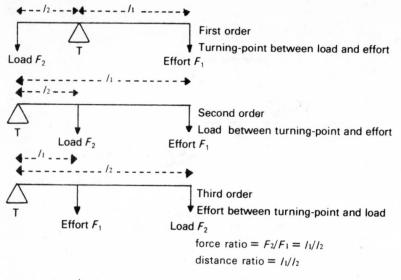

First order
Turning-point between load and effort

Second order
Load between turning-point and effort

Third order
Effort between turning-point and load

force ratio = $F_2/F_1 = l_1/l_2$

distance ratio = l_1/l_2

Lever

light-emitting diode (LED) A semi-conductor diode, made from certain materials (e.g. gallium arsenide), in which light is emitted in response to the forward-bias current. The light results from the recombination of electrons and positive holes, with a transition to a lower energy state.
See also diode, transistor.

lightning An electrical discharge in the atmosphere between charged clouds and the Earth, or between the clouds.

light-year Symbol: ly A unit of distance used in astronomy, defined as the distance that light travels through space in one year. It is approximately equal to $9.460\,5 \times 10^{15}$ metres.

limiting friction *See* friction.

linac *See* linear accelerator.

Linde process A method of liquefying gases by compression followed by expansion through a nozzle. The temperature falls by the Joule–Kelvin effect. The cooled gas is used to reduce the temperature of the compressed gas. Eventually the temperature falls below the boiling point and the gas liquefies.

The gas must start this process below its inversion temperature. Hydrogen can be liquefied by this method if it is first cooled below its inversion temperature using liquid air. Helium is cooled below its inversion temperature by using liquid hydrogen boiling under reduced pressure. *See also* cascade process.

linear accelerator (linac) A device for accelerating charged particles in a straight line. It consists of a series of cylindrical electrodes of increasing length, alternate ones of which are connected together and between which an alternating voltage is applied. A charged particle is accelerated to enter the first cylinder. By the time it emerges the voltages of the cylinders have reversed, and it is now repelled from the first cylinder and attracted towards the second, so accelerating and gaining energy each time it crosses a gap between cylinders. The cylinders have to be made progressively longer as the particles speed up so that the particles arrive at the gaps at the correct point in the voltage cycle.

A more advanced type of linear accelerator for electrons and protons uses a travelling radiofrequency electromagnetic wave in a wave guide. The particles are carried by the electric component of the wave. *See also* tandem generator.

linear charge density *See* charge density.

linear expansivity *See* expansivity.

linear momentum *See* momentum.

linear momentum, conservation of *See* constant (linear) momentum (law of).

line defect *See* defect.

lines of force Imaginary lines drawn to represent a force field. They show the direction of the field at each point; their closeness represents the field strength (the closer the lines, the greater the strength). In the case of magnetism, a line of force shows the path a free N-pole would take in the field.

Lines of force are a useful convention for showing fields; in magnetic fields they are the original basis of magnetic flux. They have, however, no real existence.

line spectrum A spectrum composed of a number of discrete lines corresponding to single wavelengths of emitted or absorbed radiation. Line spectra are produced by atoms or simple (monatomic) ions in gases. Each line corresponds to a change in electron orbit, with emission or absorption of radiation. *See also* spectrum.

line width The width of a spectral line in wavelength terms. The usual measure is the half-width of the line. *See* monochromatic radiation.

liquefaction A change of state to a liquid. The term is often used for the conversion of gases to their liquid state.

liquid *See* states of matter.

liquid barometer *See* barometer.

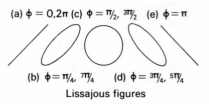

(a) $\phi = 0, 2\pi$ (c) $\phi = \pi/2, 3\pi/2$ (e) $\phi = \pi$

(b) $\phi = \pi/4, 7\pi/4$ (d) $\phi = 3\pi/4, 5\pi/4$

Lissajous figures

Lissajous' figures Patterns obtained by combining two simple harmonic motions in different directions. They can be demonstrated with an oscilloscope, deflecting the spot with one oscillating signal along one axis and with another signal along the other axis. A variety of patterns are produced depending on the frequencies and phase differences.

litre Symbol: l A unit of volume defined as 10^{-3} metre^3. The name is not recommended for precise measurements. Formerly, the litre was defined as the volume of one kilogram of pure water at 4°C and standard pressure. On this definition, $1 \, l = 1000.028 \, cm^3$.

Lloyd's mirror An arrangement in which light meets a mirror at grazing incidence and is reflected to interfere with the direct radiation. Interference of the reflected light with the direct light produces fringes. The device is effectively a two-source arrangement. *See also* interference.

logic gate An electronic circuit that can be used to perform simple logical operations. Examples of such operations are 'and', 'either — or', 'not', 'neither — nor', etc. Logic gates operate on high or low input and output voltages. Binary logic circuits, those that switch between two voltage levels (high and low), are widely used in digital computers. The *inverter gate* or *NOT gate* simply changes a high input to a low output and vice versa. In its simplest form, the *AND gate* has two inputs and

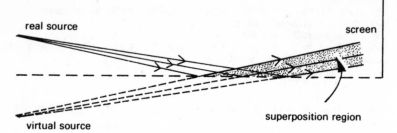

real source

screen

virtual source

superposition region

Lloyd's mirror

one output. The output is high if and only if both inputs are high. The *NAND gate* (not and) is similar, but with the output negated; that is a low output if and only if both inputs are high. The *OR gate* has a high output if one or more of the inputs are high. The *exclusive OR gate* has a high input only if one of the inputs, but not more than one, is high. The *NOR gate* has a high output only if all the inputs are low. Logic gates are constructed using transistors, but in a circuit diagram they are often shown by symbols that denote only their logical functions.
See also multivibrator, bistable circuit.

longitudinal wave A wave motion in which the vibrations in the medium are in the same direction as the direction of energy transfer. Sound waves are an example of longitudinal waves. *Compare* transverse wave.

long sight *See* hyperopia.

Lorentz–Fitzgerald contraction A reduction in the length of a body moving with a velocity *v* relative to an observer, as compared with the length of an identical object at rest relative to the observer. The object is supposed to contract by a factor $\sqrt{(1 - v^2/c^2)}$, *c* being the speed of light. The contraction was postulated to account for the negative result of the Michelson–Morley experiment in classical terms. The idea behind it was that the electromagnetic forces holding atoms together were modified by motion through the ether. The idea was made superfluous (along with the concept of the ether) by the theory of relativity, which supplied an alternative explanation of the Michelson–Morley experiment.

Lorentz force An aspect of the motor effect; the force on a charge *Q*, moving at velocity *v* across a magnetic field *B*.
$$f = BQv\sin\theta$$
θ is the angle between *v* and *B*.

loudness The sensation that a sound produces in a listener corresponding to its intensity (i.e. the energy flow per unit area). For a given frequency, the greater the intensity, the greater the loudness. Loudness also depends on the frequency of the sound. *See* phon.

loudspeaker A device for converting an electrical signal into sound. Most loudspeakers consist of a small coil, through which the signal flows, attached to the centre of a diaphragm or cone. The coil is free to move in an annular gap between the poles of a magnet. The changing current in the coil causes it to vibrate in the magnetic field and the cone sends out these vibrations as sound waves.

low frequency (LF) A radio frequency in the range between 300 kHz and 30 kHz (wavelength 1 km–10 km).

low tension (L.T.) Low voltage.

L.T. Low tension (voltage).

lumen Symbol: lm The SI unit of luminous flux, equal to the luminous flux emitted by a point source of one candela in a solid angle of one steradian. 1 lm = 1 cd sr.

luminance Symbol: L_v A measure of the brightness of an extended source (one that can not be considered a point). In a given direction, it is the luminous intensity per unit area projected at right angles to the direction. The unit is the candela per square metre (cd m^{-2}).

luminescence The emission of radiation from a substance in which the particles have absorbed energy and gone into excited states. They then return to lower energy states with the emission of electromagnetic radiation. If the luminescence persists after the source of excitation is removed it is called phosphorescence; if not, it is called fluorescence.

The excitation of the particles may occur by a variety of mechanisms. Absorption of other electromagnetic radiation gives *photoluminescence*. If the original excited states are produced by bombardment with electrons the phenomenon is *electroluminescence*. *Chemiluminescence* is luminescence produced by chemical reactions. *Bioluminescence* occurs in natural systems, e.g. glowworms and fireflies. *Radioluminescence* occurs in radioactive materials. Luminescence produced in materials by friction is called *triboluminescence*. *See also* fluorescence, phosphorescence.

luminosity Brightness; luminous intensity.

luminous exitance *See* exitance.

luminous flux Symbol: Φ_v The rate of flow of energy of visible radiation. It is the radiant flux corrected for the fact that the sensitivity of the eye is different for different wavelengths. The unit is the lumen (lm).

luminous intensity Symbol: I_v The luminous flux from a point source per unit solid angle. The unit is the candela (cd).

lunar eclipse *See* eclipse.

lunar time *See* time, year.

lux Symbol: lx The SI unit of illumination, equal to the illumination produced by a luminous flux of one lumen falling on a surface of one square metre. 1 lx = 1 lm m^{-2}.

Lyman series A series of lines in the ultraviolet spectrum emitted by excited hydrogen atoms. They correspond to the atomic electrons falling into the lowest energy level and emitting energy as radiation. The wavelength (λ) of the radiation in the Lyman series is given by $1/\lambda = R(1/1^2 - 1/n^2)$, where n is an integer and R is the Rydberg constant. *See also* spectral series.

M

machine A device for transmitting power between one place and another. The user applies a force (the effort) to the machine; the machine applies a force to a load of some kind. These two forces need not be the same. In fact the purpose of the machine may well be to overcome a large load with a small effort. For any machine this relationship is measured by the *force ratio* (or mechanical advantage): this is the force applied by the machine (load, F_2) divided by the force applied by the user (effort, F_1).

The work done by the machine cannot exceed the work done to the machine. Therefore, for a 100% efficient machine:

if $F_2 > F_1$ then $s_2 < s_1$

and if $F_2 < F_1$ then $s_2 > s_1$.

Here s_2 and s_1 are the distances moved by F_2 and F_1 in a given time.

The relationship between s_1 and s_2 in a given case is measured by the distance ratio (or velocity ratio). This is the distance moved by the effort (s_1) divided by the distance moved by the load (s_2). Neither distance ratio nor force ratio have a unit; neither has a standard symbol. *See also* efficiency, hydraulic press, inclined plane, lever, pulley, screw.

mach number The ratio of the speed of a moving object (e.g. a high-speed aircraft) to the speed of sound in the air or other medium through which the object is travelling. An aircraft passes through the 'sound barrier' as the mach number exceeds one; at this speed the air resistance increases sharply.

magnetic bottle A magnetic force field used to contain charged particles as in nuclear fusion experiments (where it contains the plasma). The plasma temperature may exceed 100 million degrees Celsius and any material container would vaporize instantly. The temperature is so high that all the electrons are stripped from the atoms and the matter consists entirely of charged particles, which can be restrained by a suitably arranged magnetic field.

magnetic circuit A closed path formed by magnetic lines of force. A simple example is given by a horseshoe magnet; its magnetic circuit is improved by the use of a keeper.

The concept is very important in electrical engineering; it compares directly to an electric circuit. Magnetomotive force F (unit — the ampereturn, At) compares to e.m.f.; magnetic flux Φ (unit — the weber, Wb) is the analogue of current; reluctance R (unit — ampere per weber, A Wb^{-1}) is like resistance.

magnetic compass *See* compass.

magnetic constant *See* permeability.

magnetic declination *See* declination.

magnetic dip *See* inclination.

magnetic dipole A pair of north-seeking and south-seeking magnetic poles a distance apart, as in a bar magnet. A loop carrying an electric current also acts as a magnetic dipole as do many atoms.

magnetic dipole moment *See* magnetic moment.

magnetic domain *See* domain.

magnetic elements The Earth's magnetism at any point is defined by three magnetic elements, which give the strength and direction of the field at that point. They are the horizontal component of the field and the angles of inclination and declination. The horizontal intensity can be measured by an Earth inductor, the inclination by an inclinometer, and the declination by a compass. In Britain the values are:
horizontal component, about 1.88×10^{-5} tesla
declination, about 9.8° West
inclination, about 66.7° North
The elements change from place to place and with time.
See also magnetic variation.

magnetic equator *See* isoclinic line.

magnetic field A region in which a magnetic force can be observed; i.e. a small magnet or a small loop of wire carrying a current will experience a force. The strength and direction of a field can be represented by lines of force. The number of lines per unit cross-sectional area is a measure of magnetic field strength — the magnetic flux density B.

magnetic field strength (magnetic intensity) Symbol: H The force that would be exerted on a unit N-pole at a given point in a magnetic field. The unit is the ampere per metre (A m^{-1}).

magnetic flux Symbol: Φ The strength of a magnetic field through an area, based on the idea of the number of lines of force passing through the area. It is given by the product of magnetic flux density (B) and area. The unit is the weber (Wb).

magnetic flux density (magnetic induction) Symbol: B The flux per unit perpendicular area of a magnetic field; it is sometimes thought of as the number of lines of force per unit area. It is defined by the effect on a current-carrying conductor in the field. For a field B:

$$B = F/Qv$$

where F is the force on a charge Q moving perpendicular to the field with velocity v.
The unit of magnetic flux density is the tesla (T). It is equivalent to the weber per square metre (Wb m^{-2}) since

$$B = \Phi A$$

where Φ is magnetic flux and A is area. In SI the relationship between magnetic flux density and magnetic field strength is:

$$B = \mu_r\mu_0 H$$

where μ_r is the relative permeability of the medium and μ_0 the permeability of free space.

magnetic focusing The use of shaped magnetic fields to focus beams of charged particles. Applications of the technique are found in cathode-ray tubes, accelerators, and electron microscopes.

magnetic hysteresis *See* hysteresis.

magnetic induction 1. The magnetization of a substance by an external magnetic field. If a sample of magnetic material is placed near a strong magnet N- and S-poles are induced in it. The sample will then act as an induced magnet and be attracted to the other. *See also* induction.
2. *See* magnetic flux density.

magnetic meridian An imaginary line on the Earth's surface joining the two magnetic poles, and passing through the observer's position. It is the line along which a compass comes to rest at that point. The angle between the magnetic and geographic meridians is the declination at that point, one of the three magnetic elements.
See also Earth's magnetism.

magnetic moment (magnetic dipole moment) Symbol: m A measure of the strength of a magnet or current-carrying coil. It relates to the turning effect (moment) on it when in a given field.
It is defined as the torque T observed in a unit field at 90° to the magnetic axis:

$$m = T/B$$

The units are ampere metres-squared (A m^2). For a coil with N turns and area A carrying a current I:

$$m = NIA$$

In this case m is often called the *electromagnetic moment* of the coil.

magnetic monopole An elementary particle postulated but not discovered, equivalent to an isolated N- or S-pole. The monopole was postulated by analogy with the proton to provide symmetry between electricity and magnetism.

magnetic permeability *See* permeability.

magnetic pole *See* poles, magnetic.

magnetic pole strength *See* pole strength.

magnetic potential *See* magnetomotive force.

magnetic resistance *See* reluctance.

magnetic susceptibility *See* susceptibility.

magnetic variation 1. Various changes with time of the magnetic elements at a point on the Earth's surface. Secular changes are slow continuous changes. Thus the angle of declination at London has changed from zero in 1659 to 8° in 1960; the inclination at the same place has changed from 74° in 1700 to 66° in 1960. Abrupt changes in the magnetic

elements are due to magnetic storms, which are related to solar activity. *See also* annual variation.
2. Note that *variation* is still widely used in navigation instead of *declination*.

magnetism The study of the nature and cause of magnetic force fields, and how different substances are affected by them. Magnetic fields are produced by moving charge — on a large scale (as with a current in a coil, forming an *electromagnet*), or on the small scale of the moving charges in the atoms. It is generally assumed that the Earth's magnetism and that of other planets, stars, and galaxies have the same cause.

Substances may be classified on the basis of how samples interact with fields. Different types of magnetic behaviour result from the type of atom. Diamagnetism, which is common to all substances, is due to the orbital motion of electrons. Paramagnetism is due to electron spin, and a property of materials containing unpaired electrons. Ferromagnetism, the strongest effect, also involves electron spin and the alignment of magnetic moments in domains. Antiferromagnetism and ferrimagnetism are rarer effects involving antiparallel alignment of spin magnetic moments.

magnetism, terrestrial *See* Earth's magnetism.

magnetocaloric effect The change of temperature of a sample as the external field changes. In particular, energy is stored in ferromagnetic domains; as these appear or disappear, energy is absorbed or released. One use of the effect is in adiabatic demagnetization for the production of very low temperatures (close to absolute zero).

magnetohydrodynamics (MHD) The study of how magnetic fields interact with conducting fluids (e.g. plasmas or liquid metals).

magnetometer A form of compass used for comparing magnetic field strengths. The *deflection magnetometer* consists

of a very small bar magnet, pivoted so that it is free to move in a horizontal plane. A circular scale, graduated in degrees, is placed below the magnet and a long light pointer is attached over it at 90°. The magnetometer is rotated in a magnetic field, H_1, so that the reading is 360. When another field, H_2, is placed at right angles to H_1 the needle is deflected by an angle θ. Then:
$$\tan\theta = H_1/H_2$$
The *vibration magnetometer* makes use of the (simple harmonic) vibration of a magnetic needle to measure the horizontal magnetic field strength at the needle. The field strength is inversely proportional to the square of the period.

magnetomotive force (magnetic potential, m.m.f.) Symbol: F The integral of the magnetic field strength around a closed path; i.e.
$$Hdl$$
The unit is the ampere-turn (At). Magnetomotive force is analogous to electromotive force in electric circuits. *See also* magnetic circuit.

magneto-optical effects Changes of the optical behaviour of matter when the external field changes. Examples are the Faraday effect and the Kerr effect.

magnetoresistive effect The change of electrical resistance of a sample when the external field changes. The Hall effect is related to this.

magnetostatics A development of magnetic theory analogous to electrostatics but based on magnetic poles rather than electric charges. Although free poles are not known, the results have some validity.

magnetostriction The change in length (and section area) of a ferromagnetic rod with change in external field. The effect is the result of domain boundary changes. Such a rod will vibrate longitudinally in an alternating field (particularly at the natural frequency); magnetostriction is a common technique for producing ultrasonic radiation.

magnet, permanent *See* permanent magnet.

magnetron A type of electron tube used for producing microwaves. A steady magnetic field is applied from outside the tube. The electrons emitted from a hot cathode circle under the influence of the field and generate electromagnetic oscillations in resonant cavities in the surrounding anode.

magnification Symbol: m The extent to which an optical system changes the apparent size of an object; i.e. the image size in comparison with the object size. It is measured by the heights perpendicular to the axis: $m = y/x$, where y is the image height and x the object height. It also depends on the distances of image and object from the centre of the system: $m = v/u$, where v is the image distance and u the object distance. If the image is smaller than the object the value of m is less than 1; a negative value of m means that the image (or object) is virtual. This is for the real is positive, virtual is negative sign convention; in the New Cartesian convention a negative value of m denotes an inverted image. *See also* angular magnification.

magnifying glass *See* microscope, simple.

magnifying power *See* angular magnification.

majority carrier *See* semiconductor.

make-and-break An arrangement by which a circuit is alternately opened and closed; for instance by movement of an armature. *See* electric bell.

manometer A device for measuring pressure. A simple type is a U-shaped glass tube containing mercury or other liquid. The pressure difference between the arms of the tube is indicated by the difference in heights of the liquid.

maser (*m*icrowave *a*mplification by *s*timulated *e*mission of *r*adiation) A device for producing an intense source of coherent microwave radiation. †Masers, like lasers, operate by population inversion and stimulated emission.

mass Symbol: m A measure of the quantity of matter in an object. Mass is determined in two ways: the *inertial mass* of a body determines its tendency to resist change in motion; the *gravitational mass* determines its gravitational attraction for other masses. The SI unit of mass is the kilogram. *See also* gravitational mass, inertial mass, weight.

mass, centre of *See* centre of mass.

mass defect The mass equivalent of the binding energy of a nucleus. *See* binding energy.

mass–energy equation The equation $E = mc^2$, where E is the total energy (rest mass energy + kinetic energy + potential energy) of a mass m, c being the velocity of light. The equation is a consequence of Einstein's Special theory of relativity; mass is a form of energy and energy also has mass. Conversion of rest-mass energy into kinetic energy is the source of power in radioactive substances and the basis of nuclear-power generation.

mass number *See* nucleon number.

mass spectrograph *See* mass spectrometer.

mass spectrometer An instrument for producing ions in a gas and analysing them according to their charge/mass ratio. The earliest experiments by J. J. Thomson used a stream of positive ions from a discharge tube, which were deflected by parallel electric and magnetic fields at right angles to the beam. Each type of ion formed a parabolic trace on a photographic plate (a *mass spectrograph*).
In modern instruments, the ions are produced by ionizing the gas with electrons from an electron gun. The positive ions are accelerated out of this

ion source into a high-vacuum region. Here, the stream of ions is deflected and focused by a combination of electric and magnetic fields, which can be varied so that different types of ion fall on a detector. In this way, the ions can be analysed according to their mass, giving a *mass spectrum* of the material. Mass spectrometers are used for accurate measurements of atomic weights, for analysis of isotope abundance, and for chemical identification of compounds and mixtures.

mass spectrum *See* mass spectrometer.

matter In general, anything that has 'substance'; a specialized form of energy with a finite rest mass, distinguished from electromagnetic radiation. The term is, perhaps, best avoided in physics.

maximum-and-minimum thermometer A thermometer designed to record the maximum and minimum temperatures during a period of time. It usually consists of a graduated capillary tube at the base of which is a bulb containing ethanol. Within the capillary is a thin thread of mercury with a steel index at each end. When the temperature rises, the alcohol expands, thus forcing the mercury up the tube, carrying the upper steel index. Therefore the position of the upper index marks the maximum temperature reached, and similarly the position of the lower index, the minimum temperature attained.

maxwell Symbol: Mx A unit of magnetic flux used in the c.g.s. system. It is equal to 10^{-8} Wb.

Maxwell distribution A distribution of velocities among the particles of a gas. The curve of number of molecules against velocity has a characteristic shape that depends on temperature. The equation:
$$N = N_0 \exp{-W/RT}$$
gives the number of molecules N with energy greater than W. N_0 is the total number of molecules.

mean free path Symbol: λ **1.** The average distance travelled by the particles of a fluid between collisions. It is given by
$$\lambda = 1/\pi r^2 n$$
where r is the particle radius and n the density of particles.
2. The average distance travelled by electrons between collisions with the lattice in conduction.

mean life Symbol: T The average time for which a radioactive nucleus exists before it disintegrates. It is the reciprocal of the decay constant.

mean solar day *See* day.

mechanical advantage *See* force ratio.

mechanical equivalent of heat *See* Joule's equivalent.

mechanics The study of forces and their effect on objects. If the forces on an object or in a system cause no change of momentum the object or system is in equilibrium. Statics is the study of such cases. If the forces acting do change momentum, dynamics is used. The ideas of dynamics relate the forces to the momentum changes produced. Kinematics is the study of motion without consideration of its cause.
See also dynamics, statics, kinematics.

medium The nature of the space through which something is transmitted or propagated. It may be solid, liquid, gas, or a vacuum (although some people do not class a vacuum as a medium).

medium frequency (MF) A radio frequency in the range between 3 MHz and 0.3 MHz (wavelength 100–1000 m).

mega- Symbol: M A prefix denoting 10^6. For example, 1 megahertz (MHz) = 10^6 hertz (Hz).

megaton A unit used to express the explosive power of nuclear weapons. One megaton is an explosive power equivalent to that of one million tons of TNT.

Meissner effect The exclusion of magnetic flux from a superconductor. Superconductors are perfect diamagnetic materials. *See* diamagnetism, superconductivity.

melting (fusion) The change from the solid to the liquid state. For a pure crystalline substance this happens at a characteristic temperature — the *melting point* (or freezing point). This temperature depends on pressure, but is usually quoted at standard atmospheric pressure. The melting point of a substance is lowered by the presence of impurities.

melting-point *See* melting.

meniscus The curved surface of a column of liquid in a tube or other container. Its shape depends on contact angle. *See also* capillary action.

meniscus lens *See* concavo-convex.

mercury barometer *See* barometer.

mercury cell A voltaic or electrolytic cell in which one or both of the electrodes consists of mercury or an amalgam. Amalgam electrodes are used in the Daniell cell and the Weston cadmium cell.

mercury thermometer A liquid-in-glass thermometer that uses mercury as its working substance. It consists of a thin capillary tube graduated in degrees, at the base of which is a bulb containing mercury. When the temperature rises, the mercury expands up the tube. The level indicates the temperature.

meridian, magnetic *See* magnetic meridian.

mesons Elementary particles that are more massive than electrons but lighter than protons and neutrons. Mesons are thought to be involved in the exchange forces between nucleons in the nucleus. They are a subclass of the hadrons. *See* elementary particles.

metastable state A condition of a system or body in which it appears to be in stable equilibrium but, if disturbed, can settle into a lower energy state. For example, supercooled water is liquid at below 0°C (at standard pressure). When a small crystal of ice or dust (for example) is introduced rapid freezing occurs.

metre Symbol: m The SI base unit of length, defined as the length equal to 1 650 763.73 wavelengths in a vacuum corresponding to the transition between the levels $2p_{10}$ and $5d_5$ of the krypton-86 atom.

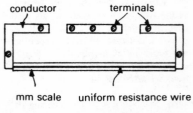

conductor   terminals

mm scale   uniform resistance wire

Metre bridge

metre bridge A metre length of uniform resistance wire mounted above a scale marked in millimetres, with suitable terminal points added to make the device adaptable either as a Wheatstone bridge or a potentiometer. As a Wheatstone bridge the wire acts as two arms of the bridge resistors; and as a potentiometer the potential drop along the wire is proportional to the length of wire. *See also* Wheatstone bridge, potentiometer.

metric system A system of units based on the metre and the kilogram and using multiples and submultiples of 10. SI units, c.g.s. units, and m.k.s. units are all scientific metric systems of units.

metric ton *See* tonne.

MF *See* medium frequency.

MHD *See* magnetohydrodynamics.

mho *See* siemens.

Michelson–Morley experiment A famous experiment carried out in 1887 in an attempt to detect the ether, the medium that was supposed to be necessary for the transmission of electromagnetic waves in free space.

In the experiment, two light beams were combined to produce interference patterns after travelling for short equal distances perpendicular to each other. The apparatus was then turned through 90° and the two interference patterns were compared to see if there had been a shift of the fringes. If light has a velocity relative to the ether and there is an ether 'wind' as the Earth moves through space, then the times of travel of the two beams would change, resulting in a fringe shift. No shift was detected, not even when the experiment was repeated six months later when the ether wind would have reversed direction. *See also* relativity.

micro- Symbol: μ A prefix denoting 10^{-6}. For example, 1 micrometre (μm) = 10^{-6} metre (m).

micron Symbol: μm A unit of length equal to 10^{-6} metre.

microphone A device for converting sound energy into electrical energy. The pressure variations of the sound wave are converted into variations in an electrical signal. There are various methods of doing this. A *ribbon microphone* has a thin metal ribbon held between the poles of a magnet. The sound waves vibrate this and a varying e.m.f. is induced, which can be amplified. A *moving-coil microphone* has a diaphragm connected to a small coil, which moves in a stationary magnetic field. An e.m.f. is induced in the coil. A *crystal microphone* depends for its action on the piezoelectric effect; an e.m.f. is induced by stress produced by sound waves hitting a suitable crystal of quartz or Rochelle salt. A *carbon microphone* has two blocks of carbon in contact. Pressure changes produced by the sound waves cause corresponding

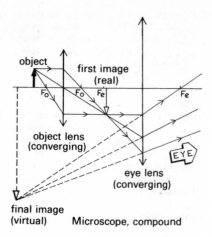

Microscope, compound

changes in the resistance of the carbon blocks.

microscope, compound A device for producing large images of close small objects with a combination of lenses. In the normal two-lens microscope each lens is converging. In practice, except in very cheap instruments, both object lens and eye lens are replaced by compound lens systems. Lighting is extremely important; light is directed onto the object by a mirror or condenser lens.

The magnifying power of the compound microscope is the product of the linear magnifications produced by objective and eyepiece.

See also electron microscope.

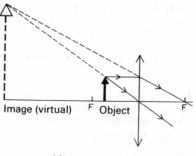

Microscope, simple

microscope, simple The magnifying glass, or simple magnifier. The small object is viewed between the lens and its focal point; an upright virtual image is obtained. In this case the image is placed at the near point of the eye — the magnifying power is then given by $(D/f) + 1$, where D is the near-point distance and f the lens focal distance. The lens can also be used to give an image at infinity; the image is then viewed with the relaxed eye, but the magnifying power is only D/f.

microwaves A form of electromagnetic radiation, ranging in wavelength from about 1 mm (where it merges with infrared) to about 120 mm (bordering on radio waves). Microwaves are produced by various electronic devices including the klyston; they are often carried over short distances in tubes of rectangular section called *wave guides*.

Microwave communication was until recently the most efficient form of electronic telecommunication. Since the development of cheap lasers and optical fibres, telecommunication at visible wavelengths is developing rapidly. Radar systems generally use microwaves, while their ability to carry energy has a number of applications, including microwave cookers. *See also* electromagnetic spectrum.

milli- Symbol: m A prefix denoting 10^{-3}. For example, 1 millimetre (mm) = 10^{-3} metre (m).

milliammeter. An instrument for measuring electric currents of a few milliamperes. *See also* ammeter, galvanometer.

minority carrier *See* semiconductor.

minute 1. A unit of time equal to 60 seconds.
2. Symbol: ' A unit of angle; 1/60 of a degree.

mirage An optical illusion caused by total internal reflection in the air. It is common when the ground is heated by the Sun so that the air in contact with the ground is warmer than the air above. Light from an object above the ground (including the sky) is refracted and totally internally reflected by the warm air to produce an image. Mirages are also sometimes observed next to very warm vertical surfaces.

mirror (reflector) An optical component that reflects incident rays. A mirror is best described by its effect on a parallel beam. Plane mirrors reflect it parallel; converging mirrors reflect it convergent; diverging mirrors make it diverge. Focal distance is the usual measure of the power of a mirror to converge a parallel beam. Parallel rays are reflected by converging and diverging mirrors as shown in the diagram. Strictly such focusing occurs only with rays parallel and close to the principal axis. With a parabolic mirror, wide parallel beams focus at the focal point. Rays from an object will be reflected so that they appear to come from somewhere else — the image. Two examples are given here to show how ray diagrams are drawn.

For a diverging reflector, a virtual image appears at F when the incident rays come from infinity; radiation from an object anywhere between infinity and P will form a virtual image between F and P.

Object distance (u) and image distance (v) relate to the focal distance (f) of the reflector as follows:
$1/f = 1/v + 1/u$ (real is positive, virtual is negative sign convention)
The formula applies to real or virtual focal points, objects, and images. All the above can be adapted for use with reflectors of non-visible radiations.

mirror image A shape that is identical to another except that its structure is reversed as if viewed in a mirror, so that the two cannot be superimposed. For example, a left hand is the mirror image of a right hand.

m.k.s. system A system of units based on the metre, the kilogram, and the second. It formed the basis for SI units.

m.k.s. system

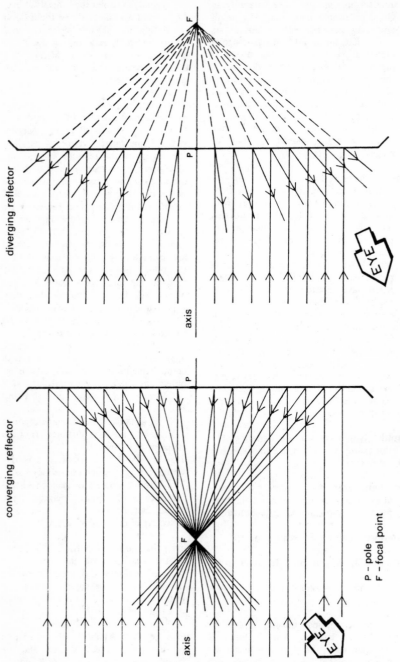

diverging reflector

converging reflector

axis

axis

EYE

EYE

P – pole
F – focal point

Paths of light reflected from converging and diverging mirrors

120

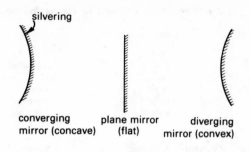

converging
mirror (concave)

plane mirror
(flat)

diverging
mirror (convex)

Types of mirror

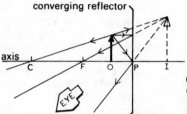

C – centre of curvature
Object 0 is between F and P
Image I is virtual, upright, magnified,
and behind the mirror

Formation of a virtual image
by a converging mirror

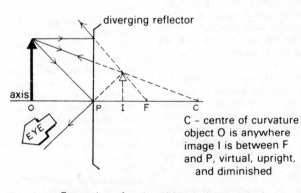

C – centre of curvature
object 0 is anywhere
image I is between F
and P, virtual, upright,
and diminished

Formation of a virtual image by a real
object at a diverging mirror

121

Object and image positions for converging mirror

Object position	Image nature	Image position
infinity	real	F
beyond C	real	between F and C
C	real	C
between C and F	real	beyond C
F	virtual	infinity
between F and P	virtual	between infinity and P

mmHg (millimetre of mercury) A former unit of pressure defined as the pressure that will support a column of mercury one millimetre high under specified conditions. It is equal to 133.322 4 Pa. It is equivalent to the torr.

moderator Material used in the cores of some nuclear reactors to slow down fast-moving neutrons so that they will be more easily captured by fissile nuclei.

The most effective speeds are about two thousand metres per second. Such neutrons are called *thermal neutrons* because the distribution of their speeds is such that they are in thermal equilibrium with (i.e. at the same temperature as) the surrounding materials. To achieve this slowing down in as few collisions as possible, moderators are substances of low atomic weight. Carbon, in the form of graphite, was used in early experimental reactors and is still used in Magnox reactors. Water — in some reactor designs, heavy water — is used and may also serve as the coolant. Paraffin wax and beryllium are also suitable moderators.

modulation The superimposition of a (video or audio) signal onto a carrier wave so that the information contained in the signal can be transmitted with the carrier wave. *See* amplitude modulation, frequency modulation, phase modulation. *See also* carrier wave, demodulation.

modulus 1. The magnitude of a vector quantity.
2. The absolute value of a number or quantity. For example, the modulus of –4 is 4.
3. A ratio that defines some property of a material or of some other system. *See* elastic modulus.

moiré fringes The pattern obtained when two regular sets of lines or points overlap. The effect can be seen through folds in curtain netting. Moiré patterns can be used as models of interference patterns. Another application is in comparing two diffraction gratings by superimposing them and observing the moiré pattern produced.

molar 1. Denoting a physical quantity divided by the amount of substance. In almost all cases the amount of substance will be in moles. For example, volume (V) divided by the number of moles (n) is molar volume $V_m = V/n$.
2. A *molar solution* contains one mole of solute per cubic decimeter of solvent.

molar thermal capacity Symbol: C_m The thermal capacity per unit amount of substance; i.e. the energy required to raise unit amount of substance (1 mol) by unit temperature (1 K). It is measured in $J\,mol^{-1}\,K^{-1}$. For a gas, it is common to specify two *principal* molar thermal capacities: one measured at constant pressure and the other measured at constant volume. The relationship between them are as for specific thermal capacities.

mole Symbol: mol The SI base unit of amount of substance, defined as the amount of substance that contains as many elementary entities as there are

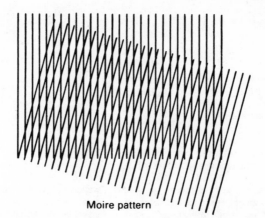

Moire pattern

atoms in 0.012 kilogram of carbon-12. The elementary entities may be atoms, molecules, ions, electrons, photons, etc., and they must be specified. One mole contains $6.022\,52 \times 10^{23}$ entities One mole of an element with relative atomic mass A has a mass of A grams (this was formerly called one *gram-atom*). One mole of a compound with relative molecular mass M has a mass of M grams (this was formerly called one *gram-molecule*).

molecular orbital *See* orbital.

molecular spectrum The absorption or emission spectrum that is characteristic of a molecule. Molecular spectra are usually band spectra.

molecular weight *See* relative molecular mass.

molecule A single atom or a group of atoms joined by chemical bonds. It is the smallest unit of a chemical compound that can have an independent existence. Water, for instance has molecules made up of two hydrogen atoms and one oxygen atom (H_2O). Ionic compounds, such as sodium chloride, do not have distinct molecules. Sodium chloride is usually written NaCl, but crystals of sodium chloride are, in fact, a regular arrangement of sodium ions (Na^+) and chloride ions (Cl^-). Similarly,

metals and certain covalently bonded solids do not have discrete molecules. In boron nitride, for instance, the covalent bonds hold the atoms together in a giant molecule.

moment The turning effect produced by a force about a point. If the point lies on the line of action of the force the moment of the force is zero. Otherwise it is the product of the force and the perpendicular distance from the point to the line of action of the force. If a number of forces are acting on a body, the resultant moment is the algebraic sum of all the individual moments. For a body in equilibrium, the sum of the clockwise moments is equal to the sum of the anticlockwise moments (this law is sometimes called the *law of moments*). *See also* torque, couple.

moment, magnetic *See* magnetic moment.

moment of inertia Symbol: I The rotational analogue of mass. The moment of inertia of an object rotating about an axis is given by $I = \Sigma mr^2$, where m is the mass of an element distant r from the axis. Some important cases are listed in the table. *See also* radius of gyration, theorem of parallel axes.

123

Moments of inertia. The value given is that for a perpendicular axis through the centre of mass.

Object	Dimensions	Moment of inertia
thin rod	length l	$ml^2/12$
thin disc	radius r	$mr^2/2$
thin ring	radius r	mr^2
solid sphere	radius r	$2mr^2/5$

momentum, conservation of *See* constant (linear) momentum (law of).

momentum, linear Symbol: p The product of an object's mass and velocity: $p = mv$. The object's momentum cannot change unless a net outside force acts. This relates to Newton's laws and to the definition of force. It also relates to the principle of constant momentum. *See also* angular momentum.

monatomic Describing a molecule that consists of a single atom. The rare gases are monatomic gases.

monochord *See* sonometer.

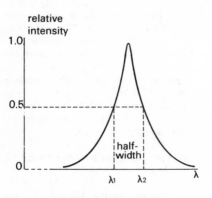

relative intensity

1.0

0.5

0

half-width

λ_1 λ_2 λ

Monochromatic light

monochromatic radiation Electromagnetic radiation of an extremely narrow range of wavelengths. (The word means 'of one colour'.) It is impossible to produce completely monochromatic radiation, although the output of some lasers is not far off. The 'lines' in line spectra produced even in the most ideal circumstances have some width in wavelength terms. The *half-width* is the measure used. It is the range of wavelengths defined in the figure, and contains almost 90% of the energy emitted. The half-width of the sharpest lines currently obtainable is about 10^{-12} m.

Simple quantum theory leads one to expect perfectly sharp lines in a line spectrum — the energies of the levels concerned appear to be exactly defined, so that $\lambda = hc/\Delta E$. However, because of the uncertainty principle, no energy level or transition *can* be defined exactly; this means that any line is naturally broadened rather than being sharp. A second broadening influence is the Doppler effect, which is relevant as the radiating particles are always in motion. Thirdly, collisions between emitting particles will broaden the emitted line. *Compare* polychromatic radiation.

monochromator A device for providing monochromatic radiation from a polychromatic source. In the case of light, for example, a prism or grating disperses the light and slits are used to select a small wavelength range.

monolithic IC *See* integrated circuit.

monostable circuit An electronic circuit, usually a multivibrator, that has one stable state but can change to another state for a time after the application of a trigger pulse. Monostable circuits can be used for generating single pulses of a fixed duration, for shortening or lengthening pulses, or to delay pulses in computer logic circuits. In · a monostable multivibrator the input pulse is fed to the base terminal of one transistor (*TR1*) and through a resistor *R* to the collector terminal of the other (*TR2*). The base of *TR2* and the collector of *TR1* are connected through a capacitor *C*. The output voltage at the collection of *TR2* is a pulse with a duration that depends on the values of *R* and *C*. See also multivibrator.

Moseley's law Lines in the X-ray spectra of elements have frequencies that depend on the proton number of the element. For a set of elements, a graph of the square root of the frequency of X-ray emission against proton number is a straight line (for spectral lines corresponding to the same transition).

sion. In less sensitive ammeters and voltmeters a pointer is used.

The torque on the coil is equal to $BIAN$, where B is magnetic flux density, I current, A coil area, and N the number of turns. The current depends on the angle of twist (θ) according to $I = k\theta/BAN$, where k is a constant of the spring or torsion wire.

moving-coil microphone †See microphone.

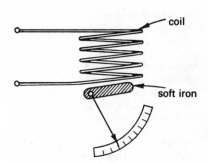

Moving-iron instrument

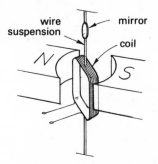

Moving-coil instrument

moving-coil instrument An electric measuring instrument that depends on the force on a small vertical rectangular coil carrying direct current in a magnetic field. The coil has a fixed cylindrical soft-iron core and is suspended between the curved pole pieces of a strong permanent magnet, designed so that the field is radial. When a current flows the coil turns; it is stopped either by torsion in a suspending wire or by a spring. The angle turned is proportional to the current. In sensitive mirror galvanometers this is measured by a small mirror fixed to the torsion-wire suspen-

moving-iron instrument An electrical measuring instrument in which the current is passed through a fixed coil. The magnetic field produced attracts a piece of soft iron on a pivot. The iron is restrained by a spring and the movement detected by a pointer. Moving-iron instruments are less sensitive than moving-coil instruments, but can be used for alternating-currents without a rectifier (the attraction does not depend on the direction of current flow). The movement of the iron is roughly proportional to the square of the current.

multimeter An electrical measuring instrument designed to measure voltage and current over a number of ranges. It is a moving-coil instrument with a switch enabling various resistors to be connected in series with the coil (for different voltage ranges) or various shunt resistances to be connected in parallel (for current ranges). Usually, an internal dry cell is incorporated so that direct measurements can also be made of resistances.

multiplication factor Symbol: k In a nuclear chain reaction, the ratio of the rate of production of neutrons to the rate of 'loss' of neutrons (by absorption and leakage).
For subcritical reactions $k < 1$
For supercritical reactions $k > 1$
For critical reactions $k = 1$

multistable circuit An electronic circuit that has more than one stable state. *See* multivibrator.

multivibrator A basic electronic circuit consisting of two transistors and other components, such as resistors and capacitors. Its main property is that it can switch its output instantaneously from one voltage level to another. Multivibrators are used to generate continuous streams of voltage pulses (square waves), to store information in the form of binary digits (0 and 1) in a computer, and to form part of a logic circuit.
The transistors are connected in the common-emitter mode with the collector of each joined through other components to the base of the other. If one transistor is in the cut-off state, that is with almost zero base current and collector current, the other is saturated, that is the collector current will not increase with collector–emitter voltage. A sufficiently high voltage pulse at the base of one transistor will change its state from cut-off to saturation and vice versa. The second transistor will then change state in the other direction. A *bistable multivibrator* circuit has two overall stable conditions and has the two transistors connected to each other through resistors only. In the *astable multivibrator* the coupling is capacitive and the circuit switches continuously from one state to another. The *monostable multivibrator* has a coupling that is both resistive and capacitive and has only one stable state. *See also* astable circuit, bistable circuit, monostable circuit.

mu-meson *See* muon.

muon A type of lepton with a mass 207 times that of the electron. There are two types, having positive or negative charge. Originally the muon was classified as a meson and called a *mu-meson*. *See also* elementary particles.

mutual inductance *See* mutual induction.

mutual induction The production of e.m.f. in one circuit by change in the current in, and therefore the magnetic field around, a second nearby circuit. For example, in a transformer, an e.m.f. is induced in one coil by a changing current in the other. Mutual induction is greatly increased by the presence of an iron core that links the two coils.
The mutual inductance (M) between two conductors is defined (in free space) by the equation $E = M(dI/dt)$, where E is the induced e.m.f. and dI/dt is the rate of change of current. The SI unit of inductance is the henry (H).
Compare self-induction.

myopia Short sight, normally caused if the distance between the lens and the back of the eyeball is too long. Light from distant objects is focused at a point in front of the retina, and thus distant objects cannot be accommodated. Only close objects can be seen clearly. The problem may be corrected by use of diverging spectacle lenses.

N

NAND gate *See* logic gate.

nano- Symbol: n A prefix denoting 10^{-9}.

For example, 1 nanometre (nm) = 10^{-9} metre (m).

natural abundance *See* abundance.

natural convection *See* convection.

natural frequency The frequency at which an object or system will vibrate freely. A free vibration occurs when there is no external periodic force and little resistance. The amplitude of free vibrations must not be too great. For instance, a pendulum swinging with small swings under its own weight moves at its natural frequency. Normally, an object's natural frequency is its fundamental frequency.

The variation of near-point distance with age

Age (years)	Distance (mm)
10	70
20	100
30	140
40	220
50	400
60	2000

near point The closest point to the eye at which an object can clearly be seen without excessive strain. The eye is then fully accommodated, and has maximum power. In the normal eye, the distance to the near point increases with age, as the eye's ability to accommodate decreases. For many purposes it is convenient to take a *standard near-point distance*, D, to be 250 mm.

near sight Short sight. *See* myopia.

Néel temperature *See* antiferromagnetism.

negative *See* charge, electric.

negative feedback *See* feedback.

negative lens A lens with a negative power. *See* diverging lens.

negative mirror A mirror with a negative power. *See* diverging mirror.

negative pole A south pole of a magnet. *See* pole.

neper Symbol: Np A unit of any ratio of powers expressed logarithmically, equal to 8.686 decibels (dB). Two powers P_1 and P_2 differ by n neper, where

$$n = \tfrac{1}{2} \ln (P_2/P_1)$$

The neper is probably most commonly used as the unit of radiation attenuation; here the radiation powers P_1 and P_2 are usually given as intensities I_1 and I_2.

Nernst calorimeter An apparatus for determining the specific thermal capacity of a metal. A piece of the metal (mass m) is wound with a coil of insulated platinum wire and suspended in an evacuated container. Aluminium foil is wrapped around the coil to reduce radiation loss. A current (I) is passed through the coil at measured voltage V for a time (t). The coil is also used as a resistance thermometer (with the heating current off) to measure the temperature rise (θ) in the metal. Then

$$IVt = (mc + C)\theta$$

where c is the specific thermal capacity of the specimen and C the thermal capacity of the coil and foil (this is found by a separate measurement).

Neumann's law *See* electromagnetic induction.

neutral Having neither negative nor positive net charge. This will be when the body is at earth potential.

neutral equilibrium Equilibrium such that if the system is disturbed a little, there is no tendency for it to move further nor to return. *See* stability.

neutral point A point where two fields in a region are equal and opposite, so that there is no resultant force. The situation is most often met in the case of magnetic fields. Thus magnetic neutral points are found near a permanent magnet in the Earth's field.

neutrino An elementary particle of zero charge and zero rest mass. Its existence, postulated in 1931 to account for the 'missing' energy and momentum in beta decay, was confirmed in 1956. There are two types of neutrino (each with a corresponding antiparticle); one type results from beta decay; the other from decay of muons. Neutrinos travel at the speed of light, and are so little affected by matter that most pass through the Earth unabsorbed, only about 1 in 10^{10} interacting with another particle. They are classified as leptons.
See also elementary particles.

neutron An elementary particle with zero charge and a rest mass of 1.674 92 $\times$ 10^{-27} kg. Neutrons are nucleons, found in all nuclides except ^{1}H. Isolated neutrons are unstable and decay with a mean life of about 12 minutes.
See also elementary particles.

neutron number Symbol: N The number of neutrons in the nucleus of an atom; i.e. the nucleon number (A) minus the proton number (Z).

New Cartesian convention *See* sign convention.

newton Symbol: N The SI unit of force, equal to the force needed to accelerate one kilogram by one metre second^{-2}. 1 N = 1 kg m s^{-2}.

Newtonian telescope The earliest type of reflecting telescope; the system is still often used. Light from the converging mirror is reflected back onto an angled plane mirror, and from there into the eyepiece. *See also* reflector.

Newton's law of cooling When a hot body is cooling in air, the rate of transfer of energy is proportional to the temperature difference between the body and its surroundings. This is strictly true only for forced convection. *See* convection.

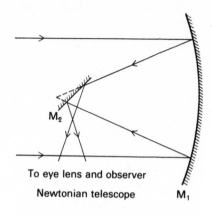

To eye lens and observer
Newtonian telescope M_1

Newton's law of universal gravitation The force of gravitational attraction between two point masses (m_1 and m_2) is proportional to the magnitude of each mass and inversely proportional to the square of the distance (r) between them. The law is often given in the form $F = Gm_1m_2/r^2$, where G is a constant of proportionality called the *gravitational constant*. The law can also be applied to bodies. It was unchallenged for over 200 years until Albert Einstein put forward his theory of relativity. *See also* relativity.

Newton's laws of motion Three laws of mechanics formulated by Sir Isaac Newton in 1687. They can be stated as:
(1) An object continues in a state of rest or constant velocity unless acted on by an external force.
(2) The resultant force acting on an object is proportional to the rate of change of momentum of the object, the change of momentum being in the same direction as the force.
(3) If one object exerts a force on another then there is an equal and opposite force (reaction) on the first object exerted by the second.
The first law was discovered by Galileo, and is both a description of inertia and a definition of zero force. The second law provides a definition of force based on the inertial property of mass. The third

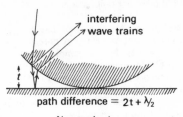

path difference = $2t + \lambda/_2$

Newton's rings

law is equivalent to the law of conservation of linear momentum.
See also reaction.

Newton's rings Interference patterns produced by placing a lens (with very large radius of curvature) on a reflecting surface and illuminating it with monochromatic light from above. Usually a plano-convex lens is used with the curved surface in contact. If the arrangement is observed from above with a microscope, light and dark rings are seen concentric with the contact point. A dark spot occurs in the centre. The radius of a bright ring is given by:
$$r^2 = a(m + \tfrac{1}{2})\lambda$$
where $m = 0, 1, 2$, etc. For the first bright ring $m = 0$, for the second $m = 1$, etc. *See also* interference.

Nichrome (*Tradename*) An alloy of about 62% nickel, 15% chromium, and 23% iron, used for making electric heating elements and resistors. It has a high resistivity and a high melting point.

nickel-iron accumulator (Edison cell, Nife or NIFE cell) A type of secondary cell in which the electrodes are formed by steel grids. The positive electrode is impregnated with a nickel-nickel hydride mixture. The negative electrode is impregnated with iron oxide. Potassium hydroxide solution forms the electrolyte. The nickel-iron cell is lighter and more durable than the lead accumulator, and it can work with higher currents. Its e.m.f. is approximately 1.3 volts.

Nicol prism A device for producing plane polarized light, consisting of a crystal of calcite cut with a 68° angle, cleaved along the optic axis, and stuck together with a thin layer of Canada balsam. The Canada balsam, which is not birefringent, has the same refractive constant for both ordinary and extraordinary rays ($n = 1.66$). The extraordinary ray passes through the prism (in calcite $n = 1.66$). However, the ordinary ray in calcite has a lower refractive constant ($n = 1.48$) than in the Canada balsam, and suffers total internal reflection at the interface. Nicol prisms are more transparent than Polaroid.

NIFE cell *See* nickel-iron accumulator.

nit Symbol: nt A unit of luminance, equal to the luminance produced by one candela per square metre (cd m^{-2}).

nodal line A line joining the nodes (positions of minimum disturbance) in an interference pattern. *See* interference.

node A point of minimum vibration in a stationary wave pattern, as near the closed end of a resonating pipe. *Compare* antinode.

noise 1. Sound composed of a random mix of different frequencies. *White noise* is a completely random mix over a wide frequency range; it has a confusing effect on the listener. *Pink noise* — random frequencies in a selected range — is often used as a background to mask other sounds.
2. A general term describing any signal that impairs the efficient working of an electronic device. There are two principal types: white noise and impulse noise. The unwanted energy of *white noise* is distributed across a wide frequency band, in the same way that the energy of white light is distributed across a band according to colour. *Impulse noise* is a consequence of one or more momentary electrical impulses. There are numerous types of white noise: thermal noise is caused by the random thermal motion of electrons

superimposed on a steady current flow; random noise can also be caused by an irregular momentary disturbance originating in an atmospheric electrical disturbance or a similar disturbance in the Sun, etc.

Noise results in undesirable sounds in a loudspeaker, which mask the desired effect, and, in television, results in 'snow', an unwanted pattern on a TV screen similar to falling snow.

non-inductive Describing a circuit component that has a low inductance. Non-inductive coils are made by doubling back the wire on itself before making the coil. The current then flows in both senses through the coil and a negligible magnetic field is produced.

non-ohmic Describing a substance or circuit component that does not obey Ohm's law; i.e. the current passed is not directly proportional to the potential difference across the circuit.

no parallax *See* parallax.

NOR gate *See* logic gate.

normal The perpendicular to a reflecting or refracting surface at the point of incidence of the ray concerned. Angles of incidence, reflection, and refraction are measured between the normal and the incident ray, reflected ray, and refracted ray respectively. A *normal ray* is one incident perpendicularly on a surface — the angle of incidence is zero.

normal adjustment An image formed by an optical system is in normal adjustment if it is in a similar viewing position to the object. The term really refers to the adjustment of the system. Thus, normal adjustment of a telescope gives the image at infinity; normal adjustment of a microscope puts the image at the viewer's near point. Other adjustments are quite possible, and may be preferred in some cases.

normal ray *See* normal.

north pole *See* poles, magnetic.

note (tone) A sound that has a single fundamental frequency. A *pure note* (or *pure tone*) has a simple harmonic waveform (approximately). Such a note can be produced by a tuning fork. Musical instruments produce complex notes, which can be analysed into a single fundamental frequency mixed with higher frequency overtones. *See* quality of sound.

NOT gate (inverter gate) *See* logic gate.

n-p-n transistor *See* transistor.

N-pole *See* poles, magnetic.

NTP Normal temperature and pressure. *See* STP.

n-type conductivity *See* semiconductor, transistor.

nuclear energy Energy derived from nuclear reactions either by the fission of heavy nuclei into lighter ones or by fusion of light nuclei into heavier ones. Nuclear energy is exploited both for weapons construction and for civil power supplies. Both fission and fusion processes have been used in weaponry but, so far, it has not proved possible to exploit fusion processes for power production. *See also* nuclear reactor, nuclear weapon.

nuclear fission The disintegration of an atomic nucleus into two or more large fragments. Fission is a form of radioactivity and, like other forms, may be spontaneous or induced by irradiation. In fission, nuclei usually disintegrate to form two fragments of roughly comparable mass number along with several neutrons. For example, the neutron-induced fission of ^{235}uranium could yield ^{90}krypton and ^{143}barium as follows:
$$^{235}_{92}U + ^{1}_{0}n \rightarrow ^{90}_{36}Kr + ^{143}_{56}Ba + 3 ^{1}_{0}n + energy$$
The energy made available will be about 30 pJ per fission. Nuclear fission is

exploited for the production of nuclear power and for nuclear weapons.

nuclear force A very strong short range attractive force acting between nucleons. Nuclear forces act over a range up to about 2×10^{-15} m and are much stronger than electromagnetic forces, so are able to overcome the electrostatic repulsion between protons in the nucleus. They are thus responsible for holding the nucleus together. See also exchange force, interaction.

nuclear fusion See fusion.

nuclear isomers A pair of isotopes of the same proton and neutron numbers, but in different quantum states. They thus have differing stability, and behave as different nuclides having different half-lives. For example, the beta decay of ^{234}Th yields two isomers of ^{234}Pa with half-lives of about 70 seconds and about 6.7 hours respectively, which may both undergo beta decay to form $^{234}_{90}$U. About 1 in 700 nuclei of the less stable isomer (that with the shorter half-life) may emit a gamma photon and make a transition to the more stable isomer.

nuclear physics The branch of physics that is concerned with nuclear structure, properties, and reactions, and their applications (e.g. in producing nuclear power or using radioisotopes).

nuclear power The use of nuclear reactions for power generation (usually by generation of electricity but nuclear powered ships use the heat generated to raise steam to power the turbines directly without conversion to electricity). The reactions concerned are usually restricted to fission and fusion reactions; some people, however, include the use of energy from radioisotope decay.

nuclear reaction A reaction involving nuclei. Such reactions include the spontaneous disintegration of a single nucleus (radioactive decay or fission); the formation of a new nucleus from two smaller nuclei (fusion); or absorption of a photon, neutron, proton, or other particle. An equation for a nuclear reaction shows the nuclides and particles involved by their symbols. For example:

$$^{210}_{82}\text{Pb} \rightarrow {}^{210}_{83}\text{Bi} + {}_{-1}^{0}e$$

represents the beta decay of lead-210 to form bismuth-210.

nuclear reactor A device in which nuclear reactions take place on a large scale. A reactor may be built for the purpose of providing useful energy, producing new nuclides, or research. Present designs are fission reactors that depend on the controlled fission of uranium. In general, they are constructed of a core containing the fuel (uranium); control rods to absorb surplus neutrons; a reflector to bounce straying neutrons back into the core; coolant to carry the energy from the core; containment — a barrier to prevent the escape of radioactive material; and shielding to protect workers from the intense radiation generated.

In *thermal reactors* a moderator is used to slow the neutrons so that a chain reaction can be sustained. *Fast reactors* do not have a moderator, and use fast neutrons. In this case the fuel has to be enriched (so that it contains more ^{235}U isotope). This type of reactor uses a blanket of natural uranium, which is converted to plutonium by neutrons. Such types are also called *converter* or *breeder reactors*.

Fusion reactors are reactors for producing energy by nuclear fusion. They are potentially important because the fuel (hydrogen) is available in unlimited quantities. However, there are immense practical problems in sustaining a controlled fusion reaction, and these reactors are still in the development stage. See also breeder reactor, †boiling-water reactor, gas-cooled reactor, pressurized water reactor.

nuclear weapon An explosive device in which the reduction in mass that provides the thermal energy results either from the fission of heavy nuclei (such as ^{238}U) in a rapidly built up chain reaction or from the fusion of light nuclei such as deuterium and tritium to form helium. This latter type of device is known as a 'hydrogen bomb'. Although the total number of nucleons in the product is the same as in the original nuclei the mass of the resulting nucleus is less. See also critical mass, mass defect.

nucleon A particle found in the nucleus of atoms; i.e. a proton or a neutron. Nucleons are classified as baryons. See also elementary particles.

nucleon number (mass number) Symbol: A The number of nucleons (protons plus neutrons) in an atomic nucleus.

nucleus, atomic The compact, comparatively massive, positively charged centre of an atom made up of one or more nucleons (protons and neutrons) around which is a cloud of electrons. The density of nuclei is about 10^{15} kg m^{-3}. The number of protons in the nucleus defines the element, being its atomic number (or proton number). The nucleon number, or atomic mass number, is the sum of the protons and neutrons. The simplest nucleus is that of a hydrogen atom, ^{1}H, being simply one proton (mass 1.67×10^{-27} kg). The most massive naturally occurring nucleus is ^{238}U of 92 protons and 146 neutrons (mass 4×10^{-25} kg, radius 9.54×10^{-15} m). Only certain combinations of protons and neutrons form stable nuclei. Others undergo spontaneous decay.
A nucleus is depicted by a symbol indicating nucleon number (mass number), proton number (atomic number), and element name. For example, $^{23}_{11}$Na represents a nucleus of sodium having 11 protons and mass 23, hence there are (23 − 11) = 12 neutrons.

nuclide A nuclear species with a given number of protons and neutrons; for example, ^{23}Na, ^{24}Na, and ^{24}Mg are all different nuclides. Thus:
$^{23}_{11}$Na has 11 protons and 12 neutrons
$^{24}_{11}$Na has 11 protons and 13 neutrons
$^{24}_{12}$Mg has 12 protons and 12 neutrons

The term is applied to the nucleus and often also to the atom.

O

object A real or apparent source of rays in an optical system, perhaps incident on a lens or a reflector. After refraction or reflection, the rays appear to come from some other place — the image.
An object need not be real. The diagram shows how the real image I, produced by the lens, becomes a virtual object O_2 when the mirror is introduced. This now gives a real image I_2. Just as with real objects, virtual objects can appear as real or virtual images.

objective The lens or combination of lenses nearest the object in an optical instrument. The nearest lens to the object in a compound objective is often called the object lens. The large converging mirror in a reflecting telescope can also be described as the objective.

object lens See objective.

object plane The plane perpendicular to the axis centred on the object.

octave The interval between a waveform and another of twice the frequency. An octave in sound corresponds to eight notes on the diatonic musical scale.

ocular See eyepiece.

oersted Symbol: Oe A unit of magnetic field strength in the c.g.s. system. It is equal to $10^3/4\pi$ A m^{-1}.

ohm Symbol: Ω The SI unit of electrical resistance, equal to a resistance that passes a current of one ampere when there is an electric potential difference of one volt across it. 1 Ω = 1 V A^{-1}. Formerly, it was defined in terms of

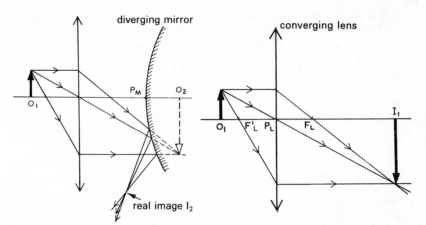

Formation of a real image involving a virtual object

Formation of a real image from a real object

the resistance of a column of mercury under specified conditions.

ohmic Describing a substance or circuit component that obeys Ohm's law.

Ohm's law The current (I) in a conductor is proportional to the potential difference (V) between its ends. This leads to $V = IR$, where R is the conductor's resistance. R is constant provided the physical conditions in the conductor are unaltered. If V is in volts and I in amperes, R is in ohms. Components that conduct in accordance with Ohm's law (e.g. metal resistors) are said to be *ohmic*.

oil-immersion objective *See* immersion objective.

omega particle *See* elementary particles.

opaque Not able to pass radiant energy. A substance that is opaque to one type of electromagnetic radiation may be transparent to another. Thus glass passes visible radiation very well, but is opaque to most thermal (infrared) radiation. *Compare* translucent.

opaque projector (episcope) An optical instrument for projecting an image of an opaque object (e.g. a diagram or picture) onto a screen. High-intensity illumination is used and the image is projected by a combination of mirrors and lenses.

open circuit *See* circuit, electrical.

opera glasses *See* binoculars.

optical activity The ability of certain crystals or compounds in solution to rotate the plane of polarization of plane polarized light. Compounds that are optically active have an asymmetric molecular structure such that their molecules can exist in left- and right-handed forms (the forms are mirror images of each other; the property of having such forms is called *chirality*). Particular forms of the compound are classified as *dextrorotatory* (right turning) or *laevorotatory* (left turning). Laevorotatory compounds rotate the plane of polarization to the left; i.e. anticlockwise as viewed facing the oncoming light. Dextrorotatory compounds rotate the plane to the right (i.e. in the opposite sense). Many naturally occurring substances (e.g. sugars) are optically active.

optical axis The principal axis of an optical system, i.e. the path of rays passing along the principal axes of the lenses or mirrors.

optical centre (optic centre) *See* pole.

optical fibre *See* fibre optics.

optical glass A type of glass with properties suitable for making lenses, prisms, etc. There are two major groups: the *crown glasses* and the *flint glasses*, differing in their density and refractive constants. Optical glasses are fairly hard yet easily polished and highly transparent to light.

optical lever A device for measuring angular displacement. A small mirror is attached to the rotating body and a narrow fixed beam of light directed onto it. The reflected beam is directed onto a screen, producing a spot of light. Movement of this spot shows small angular movements of the mirror. The angle turned through by the reflected beam is twice that turned by the mirror. The device is commonly used in torsion balances, as in the mirror galvanometer.

optical path Symbol: d The product of distance (l) travelled in a medium and the refractive constant (n) of the medium; i.e. $d = nl$. Phase difference ($\Delta\phi$) relates to path difference (Δd) thus:

$$\Delta\phi = 2\pi\Delta d/\lambda$$

optical pyrometer (disappearing-filament pyrometer) A type of pyrometer used to measure the temperature of incandescent sources. Light from the source is focused onto a tungsten filament, and the filament and the image of the source are viewed through a red filter and an eyepiece. The current in the filament is varied until this has the same brightness as the body (i.e. it cannot be seen against the background of the source). The ammeter is calibrated directly in degrees Celsius, with the assumption that the source emits black-body radiation. A correction can be made for the spectral emissivity of the source. *See also* total-radiation pyrometer.

optical rotary dispersion (ORD) The phenomenon in which the amount of rotation of plane-polarized light of an optically active substance depends on the wavelength. Plots of rotation against wavelength can be used to give information on molecular structure.

optical telescope A telescope using visible light from a distant object to produce a magnified image. *See* telescope.

optic axis The direction in a birefringent crystal along which the ordinary and extraordinary rays travel at the same speed. Uniaxial crystals have one such axis; biaxial crystals have two. *See also* birefringent crystal.

optics The study of the nature and behaviour of light and other radiations. The reflection of ultraviolet radiation and the refraction of sound (pressure) waves also follow the laws of optics. Electron diffraction and the electron microscope are branches of electron optics.
Where the wave nature of the radiation need not be considered, situations can be discussed in terms of rays. That study is traditionally called *geometrical optics. Physical optics* is the field of optics in which wave properties are important.

Optical glass

Glass	Density (kg m⁻³)	Refractive constant
crown	2500–2600	1.51–1.61
flint	2700–4800	1.53–1.75

Thus the use of lenses is part of geometrical optics, while the diffraction grating comes into physical optics.

optoelectronics A branch of electronics based on light beams rather than electric currents. The main advantage is that light frequencies are far higher than radio frequencies so that much more information can be carried in a suitably treated light beam than in an electronic (radio) signal. Optical fibres, of very cheap very transparent glass, carry the light beams, produced by semiconductor lasers and detected by photocells.

orbit The curved, usually closed, path or trajectory along which a moving object travels. Examples include the elliptical orbit of a planet in the solar system, the circular orbit of an electron in a magnetic field, and the parabolic trajectory of a projectile under gravity.

orbital 1. Pertaining to an orbit. For example, an orbital electron is one that is bonded around the nucleus of an atom.
2. According to quantum theory, the electrons in an atom do not have fixed orbits around the nucleus. Instead, there is a finite probability of finding the electron in a given volume at any distance from the nucleus. The region in space in which there is a high probability of finding an electron is an *atomic orbital*. For a hydrogen atom in its ground state the orbital is a spherical shell around the nucleus. Other types of orbital have different shapes. Similarly, in molecules, electrons move in *molecular orbitals* around the nuclei of the atoms.

order An integer (m) associated with a given interference fringe or diffraction pattern. In interference a bright fringe occurs for a path difference $m\lambda$; a dark fringe is produced if the path difference is $(m + \frac{1}{2}) \lambda$. A bright fringe is first order if it arises through a path difference of one wavelength ($m = 1$). Similarly, second order corresponds to $m = 2$, etc.

ordinary ray *See* birefringent crystal.

OR gate *See* logic gate.

oscillation A regularly repeated motion or change. *See* vibration.

oscillator A device that generates an alternating current signal of known frequency from a direct current input. The output is usually in the form of a sine wave or a square wave. The oscillator may be regarded as an amplifier supplying its own energy input, the oscillation being maintained by positive feedback to make good the energy losses. A sinusoidal oscillation can be produced by combining an inductor and a capacitor in a resonant circuit. The frequency is dependent on the values of the inductance and the capacitance. A square wave is produced in a multivibrator by the alternate charge and discharge of a capacitor, the frequency being dependent on the resistance and capacitance. *See also* crystal oscillator, multivibrator.

oscilloscope *See* cathode-ray oscilloscope.

Otto cycle The idealized reversible cycle of four operations occurring in a perfect four-stroke (spark ignition) petrol engine. These are the constant-volume temperature rise, constant-pressure expansion, constant-volume temperature fall, and constant-pressure volume decrease of a gas. The cycle returns the gas to its initial state; the efficiency with which the engine transfers chemical energy into mechanical work is the maximum attainable for this type of engine. *Compare* Carnot cycle.

overdamping *See* damping.

overhead projector A type of projector able to project, on a screen, large bright images of slides or transparent objects placed on a horizontal table. *See also* projectors.

overtone A component of a note that has

higher frequency and (usually) lower intensity than the fundamental. The overtones have frequencies that are simple multiples of the fundamental frequencies. The word is a musical term best avoided in physics. *See* partials, quality of sound.

P

pair production The production of an electron and its antiparticle (a positron) from a photon (according to the equation $E = mc^2$). The process can occur in the field of an atomic nucleus. Since the mass of an electron or positron is equivalent to 0.511 MeV, the minimum energy of a photon that can promote pair production is 1.022 MeV. Any surplus energy becomes kinetic energy of the products.

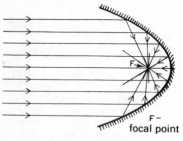
Parabolic mirror

parabolic mirror A reflector with a parabolic section. The converging type can converge wide parallel beams accurately into its focal point (and is thus used for reflection telescopes and solar power applications). On the other hand, radiation from a source at the focal point will be reflected into a parallel beam (as used in various lighting systems).

parallax The apparent difference in an object's position relative to that of another when viewed from two different places. If there is *no parallax* between two objects, they must be at effectively the same place. Often in experiments with light, no-parallax methods are used to locate images. When there is no parallax between the image and the object used to find it, the two are in the same position.

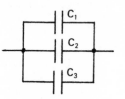
capacitors in parallel

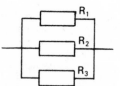
resistors in parallel

Circuits

parallel Elements in an electrical circuit are *in parallel* if connected so that the current divides between them and rejoins at the other side. The word 'shunt' is sometimes used.

For resistors in parallel, the resulting resistance R is given by:
$1/R = 1/R_1 + 1/R_2 + 1/R_3 + \ldots$
For capacitors in parallel, the capacitance of the combination is given by:
$C = C_1 + C_2 + C_3 + \ldots$
For cells in parallel, the e.m.f. is equal to the largest value of e.m.f. of all the cells. *Compare* series.

parallel axes, theorem of *See* theorem of parallel axes.

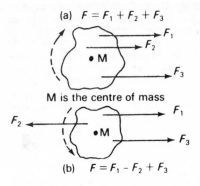

(a) $F = F_1 + F_2 + F_3$

M is the centre of mass

(b) $F = F_1 - F_2 + F_3$

Parallel forces

parallel forces When the forces on an object pass through one point, their resultant can be found by using the parallelogram of vectors. If the forces are parallel the resultant is found by addition, taking sign into account. There may also be a turning effect in such cases, which can be found by the principle of moments.

parallelogram (law) of forces *See* parallelogram of vectors.

parallelogram (law) of velocities *See* parallelogram of vectors.

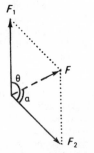

Parallelogram of vectors

parallelogram of vectors A method for finding the *resultant* of two vectors acting at a point. The two vectors are shown as two sides of a parallelogram: the resultant is the diagonal through the starting point. The technique can be used either with careful scale drawing or with trigonometry. The trigonometrical relations give:

$$F = \sqrt{(F_1^2 + F_2^2 + 2F_1F_2\cos\theta)}$$
$$\alpha = \sin^{-1}[(F_1/F)\sin\theta]$$

paramagnetism The magnetic behaviour of substances with a small susceptibility; this often varies with temperature according to Curie's law. The relative permeability of a paramagnetic substance is slightly greater than one. Paramagnetism results from unpaired electron spins. In an external field, a piece of paramagnetic material will slightly concentrate, and thus increase, the field; this is because the magnetic moments of the particles align themselves in the same direction. As a result, the sample itself will align parallel to the field. Above its Curie temperature, a ferromagnetic material will have no domains and will show only paramagnetic behaviour. *See also* magnetism.

paraxial Describing rays incident on a surface close and parallel to the axis. Only paraxial rays pass or appear to pass through the focal point of a spherical reflecting or refracting surface. *See also* mirror, parabolic mirror.

parent A nuclide that undergoes radioactive decay to another specified nuclide (the *daughter*).

parity Symbol: P A fundamental property of a system that can be thought of as the ability to be reflected in a mirror. If inverting all the signs of all the coordinates makes no change in parity, P, then P is said to be even, and to have value +1. If the procedure inverts the sign, then parity is odd and equal to -1. Parity is a quantum number and is said to be conserved in an interaction if the parity of the products (found by multiplying together their separate parities) is equal to the parity of the initial arrangement. Parity is conserved in strong interactions, but not in weak interactions, such as beta decay.

parsec Symbol: pc A unit of distance used in astronomy. A star that is one parsec away from the earth has a parallax (apparent shift), due to the Earth's movement around the Sun, of one second of arc. One parsec is approximately $3.085\ 61 \times 10^{16}$ metres.

partial eclipse See eclipse.

partial pressure See Dalton's law.

partials Components of a sound wave with frequencies above the fundamental frequency. The partials include the overtones together with non-harmonic components.

particle An abstract simplification of a real object — the mass is concentrated at the object's centre of mass; its volume is zero. Thus relational aspects can be ignored.

particle accelerator See accelerator.

particle, elementary See elementary particle.

pascal Symbol: Pa The SI unit of pressure, equal to a pressure of one newton per square metre ($1\ Pa = 1\ N\ m^{-2}$). The pascal is also the unit of stress.

Paschen series A series of lines in the infrared spectrum emitted by excited hydrogen atoms. The lines correspond to the atomic electrons falling into the third lowest energy level and emitting energy as radiation. The wavelength (λ) of the radiation in the Paschen series is given by $1/\lambda = R(1/3^2 - 1/n^2)$ where n is an integer and R is the Rydberg constant. See also spectral series.

Pauli exclusion principle See exclusion principle.

p.d. See potential difference.

peak value The maximum value attained by an alternating quantity (e.g. an alternating current).

Peltier effect The change in temperature produced at a junction between two different metals when electric charge is passed through it. If the current direction is reversed then a heating effect becomes a cooling one or vice versa. The temperature change is directly proportional to the current. Compare Seebeck effect.

pencil A narrow beam of rays from a single point. See beam.

pendulum A simple pendulum consists of a small mass oscillating to and fro at the end of a very light string. If the amplitude of oscillation is small (less than about 10°), it moves with simple harmonic motion. The period does not depend on amplitude; there is a continuous interchange of potential and kinetic energy. The period is given by
$$T = 2\pi\sqrt{(l/g)}$$
Here l is the length of the pendulum (from support to centre of the mass) and g is the acceleration of free fall.

A compound pendulum is a rigid body swinging about a point. The period depends on the moment of inertia. For small oscillations it is given by the same relationship as that of the simple pendulum with l replaced by $[\sqrt{(k^2 + h^2)}]/h$. Here, k is the radius of gyration about an axis through the centre of mass and h is the distance from the pivot to the centre of mass.

pentode A thermionic valve with three grids, one grid more than the tetrode. The extra grid, called the suppressor grid, is placed between the screen grid and the anode. It is held at cathode potential, which suppresses by repulsion the loss of secondary electrons ejected by the anode. Thus the disadvantage of the tetrode is overcome. See also grid, tetrode, thermionic valve.

penumbra Partial shadow. See shadow.

perfect gas *See* gas laws, kinetic theory.

period Symbol: T The time for one complete cycle of an oscillation, wave motion, or other regularly repeated process. It is the reciprocal of the frequency, and is related to pulsatance, or angular frequency, (ω) by $T = 2\pi/\omega$.

periodic motion Any kind of regularly repeated motion, such as the swinging of a pendulum, the orbiting of a satellite, the vibration of a source of sound, or an electromagnetic wave.

If the motion can be represented as a pure sine wave, it is a simple harmonic motion. Harmonic motions in general are given by the sum of two or more pure sine waves.

periscope An optical instrument enabling the user to see over or round an obstacle. The simplest version is the mirror periscope, which has two mirrors, each at 45° to the viewed direction. Rather more effective types use internally reflecting prisms. In submarines, telescopic lenses are fitted.

permanent gas A gas that cannot be liquefied by increasing its pressure at room temperature, but must also be cooled. Such a gas has a critical temperature below room temperature. *See* critical state.

permanent magnet A sample of a substance that retains its magnetism when the external magnetic field is removed. Permanent magnets are often made from steel, other alloys, or ferrites; they have many uses. Permanent magnetism is mainly a property of ferromagnetic materials that have high coercivity. Materials like this are often said to be magnetically 'hard'. *Compare* temporary magnetism.

permeability Symbol: μ In SI, *absolute permeability* is the ratio of the magnetic flux density (B) in a substance to the external magnetic field strength (H); i.e.
$$\mu = B/H$$
The unit is the henry per metre (H m^{-1}).

The permeability of free space (i.e. a vacuum) has a value of $4\pi \times 10^{-7}$ H m^{-1}, and is given the symbol μ_0. The *relative permeability* (μ_r) of a substance is the ratio of its absolute permeability to the permeability of free space; i.e.
$$\mu_r = \mu/\mu_0$$
Note that relative permeability, being a ratio, has no units. It is related to susceptibility (χ) by:
$$\mu_r = 1 + \chi$$

permittivity Symbol: ϵ The mutual force between two charges Q_1 and Q_2 a distance r apart is given by the equation:
$$F = Q_1 Q_2 / 4\pi\epsilon r^2$$
The constant ϵ is the permittivity of the medium. The unit is the coulomb^2 newton^{-1} metre^{-2} ($\text{C}^2 \text{N}^{-1} \text{m}^{-2}$), or farad metre^{-1} ($\text{F m}^{-1}$). The permittivity of free space, ϵ_0 (also called the *electric constant*) has the value 8.854×10^{-12} F m^{-1}. The absolute permittivity of a material is equal to $\epsilon_0\epsilon_r$, where ϵ_r is the relative permittivity. *See also* Coulomb's law.

perpetual motion Constant motion occurring without any external energy supply. Perpetual-motion machines are impossible to construct because of frictional and other forces dissipating the original energy.

perversion (lateral inversion) The effect produced by a mirror in reversing images apparently left to right. Thus writing appears backwards when reflected; the image of a person raising the right hand appears to raise the left hand.

Perversion is not of unusual significance. In the case of a plane mirror it follows directly from the fact that each point of the object produces an image point that is directly opposite it behind the mirror.

phase 1. A homogeneous part of a mixture distinguished from other parts by boundaries. A mixture of ice and water has two phases. A mixture of ice crystals and salt crystals also has two phases. A solution of salt in water is a single phase.

2. The stage in a cycle that a wave (or other periodic system) has reached at a particular time (taken from some reference point). Two waves are *in phase* if their maxima and minima coincide.

For a simple wave represented by the equation

$$y = a\sin 2\pi(ft - x/\lambda)$$

the phase of the wave is the expression $2\pi(ft - x/\lambda)$. The *phase difference* between two points distances x_1 and x_2 from the origin is $2\pi(x_1 - x_2)/\lambda$.

A more general equation for a progressive wave is

$$y = a\sin 2\pi(ft - x/\lambda - \phi)$$

Here, ϕ is the *phase constant* — the phase when t and x are zero. Two waves that are out of phase have different phase constants (they 'start' at different stages at the origin). The phase difference is $|\phi_1 - \phi_2|$. It is equal to $2\pi x/\lambda$, where x is the distance between corresponding points on the two waves. It is the *phase angle* between the two waves; the angle between two rotating vectors (phasors) representing the waves.

See also beats, wave.

phase angle *See* phase.

phase constant *See* phase.

phase difference The difference in the phases of two coherent radiations in a superposition region. If the phase difference ($\Delta\phi$) is 0, 2π, 4π, etc., the two are in phase and will interfere constructively. If $\Delta\phi$ is π, 3π, 5π, etc., the two are in anti-phase and will interfere destructively. *See also* path difference.

phase modulation A type of modulation in which a carrier wave is made to carry the information in a signal (audio or visual) by fluctuations in the phase of the carrier. The difference in phases between the modulated and the unmodulated carrier is proportional to the amplitude of the signal. The carrier remains at the same frequency. The carrier amplitude is constant. *See also* carrier wave, modulation.

phase velocity The velocity with which the phase in a travelling wave is propagated. It is equal to λ/T, where T is the period. *Compare* group velocity.

phasor A rotating vector used to represent a sinusoidally varying quantity (e.g. an alternating current). The projection of the vector on a fixed axis represents the amplitude variation with time. A phase angle between two quantities (e.g. current and voltage) is represented by the angle between their phasors.

phi particle *See* elementary particles.

phon Symbol: p A unit for measuring the loudness of sounds. Noises of the same intensity sound louder or softer, depending on the frequency. The phon is defined using a standard reference source of 1000 hertz, with which other sounds are compared. A loudness of n phons is the same as that of a standard source with an intensity of n decibels above the threshold of hearing (10^{-12} W m^{-2}).

phosphor A substance that shows luminescence or phosphorescence.

phosphorescence **1.** The absorption of energy by atoms followed by emission of electromagnetic radiation. Phosphorescence is a type of luminescence, and is distinguished from fluorescence by the fact that the emitted radiation continues for some time after the source of excitation has been removed. In phosphorescence the excited atoms have relatively long lifetimes before they make transitions to lower energy states. However, there is no defined time distinguishing phosphorescence from fluorescence.

2. In general usage the term is applied to the emission of 'cold light' — light produced without a high temperature. The name comes from the fact that white phosphorus glows slightly in the dark as a result of a chemical reaction with oxygen. The light comes from excited atoms produced directly in the

reaction — not from the heat produced. It is thus an example of *chemiluminescence*. There are also a number of biochemical examples (bioluminescence); for example, phosphorescence is sometimes seen in the sea from marine organisms, or on rotting wood from certain fungi (known as 'fox fire').
See also luminescence.

phot A unit of illumination in the c.g.s. system, equal to an illumination of one lumen per square centimetre. It is equal to 10^4 lx.

photocathode A cathode that emits electrons by the photoelectric effect.

photocell Any device for producing an electric signal from electromagnetic radiation. Originally, photocells were photoelectric cells — i.e. devices in which a current was produced between two electrodes by the photoelectric effect. The present common type depends on photoconductivity. A piece of semiconductor material is held between two contacts and a voltage applied. In the absence of radiation the current is very small; when radiation falls on the sample, its resistance is reduced and the current increases. Photoconductive cells, unlike photoelectric cells, can be used in the infrared region. Other types of photocell are photodiodes or depend on the photovoltaic effect. *See also* photoelectric cell.

photoconductivity The increase of electrical conductivity in certain materials, produced by incident electromagnetic radiation. The photons absorbed by the solid give up energy to electrons, freeing them and thus increasing the number of electrons in the conduction band.

photodiode A semiconductor diode that is sensitive to electromagnetic radiation. The p–n junction is reverse biased. When radiation falls on it, electron-hole pairs are created. Some of these additional charge carriers are immediately swept across the junction before they recombine, constituting a current that depends on the illumination. *See also* phototransistor.

photodisintegration The decay of an atomic nucleus following the absorption of electromagnetic energy (gamma-ray photons).

photoelasticity The double refraction observed in certain materials only when stressed. A number of synthetic materials are photoelastic, examples being Perspex and Cellophane. Polarized white light passed through a stressed sample shows coloured fringes relating to the stress pattern. This is a very useful engineering test technique. Glass also shows interference fringes and the method is used for locating strains in glass laboratory apparatus.

photoelectric cell A device in which an electric current is produced by electromagnetic radiation (visible or ultraviolet) by the photoelectric effect. A light-sensitive cathode (photocathode) and an anode are placed in an evacuated glass envelope. The photosurface contains material with low work function (e.g. potassium or caesium) and emits electrons when irradiated. A positive potential on the anode causes a current to flow between the electrodes and in an external circuit. The device used to be called a *photocell*. In most applications it has been superseded by the photoconductive cell. *See also* photocell.

photoelectric cell *See* photocell.

photoelectric effect The emission of electrons from a solid (or liquid) surface when it is irradiated with electromagnetic radiation. For most materials the photoelectric effect occurs with ultraviolet radiation or radiation of shorter wavelength; some materials show the effect with visible radiation.

In the photoelectric effect, the number of electrons emitted depends on the intensity of the radiation and not on its frequency. The kinetic energy of the electrons that are ejected depends on the

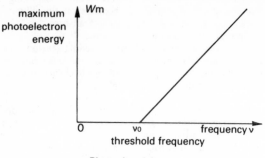

maximum
photoelectron
energy

Wm

0   v0   frequency v

threshold frequency

Photoelectricity

frequency of the radiation. This was explained, by Einstein, by the idea that electromagnetic radiation consists of streams of photons. The *photon energy* is $h\nu$, where h is the Planck constant and ν the frequency of the radiation. To remove an electron from the solid a certain minimum energy must be supplied, known as the *work function*, ϕ. Thus, there is a certain minimum threshold frequency ν_0 for radiation to eject electrons: $h\nu_0 = \phi$. If the frequency is higher than this threshold the electrons are ejected. The maximum kinetic energy (W) of the electrons is given by *Einstein's equation*: $W = h\nu - \phi$. The photoelectric effect also occurs with gases. *See* photoionization.

photoelectricity A group of phenomena in which electric effects are produced by electromagnetic radiation. *See* photoconductivity, photoelectric effect, photoionization, photovoltaic effect.

photoelectrons Electrons ejected from a solid, liquid, or gas by the photoelectric effect or by photoionization.

photoemission The emission of photoelectrons by the photoelectric effect or by photoionization.

photography The production of a permanent record of an image on suitable paper or film. Normally the image is produced by the optical system

of some type of camera on the light-sensitive surface of an emulsion. This is a layer of a silver salt on glass, paper, or film: the molecules of the compound are decomposed to silver atoms by the light energy, forming a latent image. This is developed and fixed to make the permanent photograph.

Photographic emulsions are 'sensitive' to electromagnetic radiations of wavelengths less than around 1.5 μm. They are also sensitive to other energetic ionizing radiations, and are important in X-ray investigations and in nuclear physics. Colour photography involves mixing specified dyes into the silver halide–gelatine emulsion.

photoionization The ionization of atoms or molecules by electromagnetic radiation. Photons absorbed by an atom may have sufficient photon energy to free an electron from its attraction by the nucleus. The process is $M + h\nu \rightarrow M^+ + e^-$. As in the photoelectric effect, the radiation must have a certain minimum threshold frequency. The energy of the photoelectrons ejected is given by $W = h\nu - I$, where I is the ionization potential of the atom or molecule.

photoluminescence *See* luminescence.

photometer An instrument for determining illumination or luminous intensity by comparisons made with the eye. There are various types. The general principle is to compare a source with a standard source. The two sources each illuminate a screen, and the positions of the sources are adjusted until both give equal illumination. The illumination (E) is related to luminous intensity by $E = (I \cos i)/d^2$, where I is the luminous intensity, d the distance from the source to surface, and i the angle of incidence.

photometry The branch of physics concerned with measuring intensity, illumination, etc., for visual radiation.
There are two basic types of measurement. 'Luminous' quantities, such as luminous flux, luminous intensity, and illumination, which are based on measurements made with the eye (e.g. by comparing the illumination of two surfaces). They thus depend on the sensitivity of the eye to different wavelengths. 'Radiant' quantities (e.g. radiant flux, radiance, and irradiance) depend on absolute measures of energy flow made by photocells. *See also* Lambert's laws.

photomultiplier A device in which electrons originally emitted from a photocathode initiate a cascade of electrons by secondary emission in an electron multiplier. It is much more sensitive as a radiation detector than a single photoelectric cell. *See also* photoelectric cell.

photon A 'particle' of electromagnetic radiation. Although most aspects of the behaviour of electromagnetic radiation appear to show that it consists of waves, a few do not. The photoelectric effect was the first of these to be discovered, about a century ago. It seems that light and other similar radiations travel in a manner that can show wave behaviour in some situations and particle behaviour in others.
The energy (W) carried by a photon is related to the frequency (ν) by the Planck constant (h): $W = h\nu$. Its momentum p is given by $p = h/\lambda$ (where λ is the wavelength) and by $p = W/c$. The speed of photons is the speed of light (c); the rest mass is zero.

photopic vision Vision at normal levels of light intensity (as in normal daylight). Under such conditions the cones in the retina are the main receptors. *Compare* scotopic vision.

phototransistor A transistor that is sensitive to electromagnetic radiation. When radiation falls on the emitter part of the device, more charge carriers are produced in the base region and the collector current increases. A phototransistor acts like a photodiode with an amplifier. *See also* photodiode.

photovoltaic effect The effect in which irradiation of a p-n junction or the junction of a metal and a semiconductor by electromagnetic radiation (ultraviolet to infrared) generates an e.m.f., which can be used to deliver power to an external circuit. A solar cell consists of such a p-n junction. *See also* semiconductor.

photovoltaic effect The production of an e.m.f. between two layers of different materials by incident electromagnetic radiation. *See also* phototransistor.

physical optics *See* optics.

physisorption *See* adsorption.

pico- Symbol: p A prefix denoting 10^{-12}. For example, 1 picofarad (pF) = 10^{-12} farad (F).

piezoelectric effect The generation of a small potential difference across certain materials when they are subjected to a stress. The effect is used in devices for pressure and vibration measurement, in gas lighters, in strain gauges, and in oscillators. *Compare* magnetostriction.

pi-meson *See* pion.

pin-hole camera A box into which light can enter only through a very small hole. Because light rays travel in straight lines in a given medium (air in this case), an inverted image is formed on the wall of the box opposite the pin hole. The wall may carry a photographic plate, or be made of a diffusing material such as tissue paper or ground glass.

Because the hole is small, the image is very faint — a long exposure is needed to make a photograph. However the image is in sharp focus, whatever the distance of the object. This is because the rays pass through the hole without deviation. Unlike the lens of a lens camera, the pin hole gives no image aberrations. If the hole is made larger, the image becomes brighter — but more blurred. A large hole can be thought of as a set of pin-holes — each pin hole gives a faint image, but the images are in slightly different positions.

pink noise See noise.

pion (pi-meson) A type of meson. There are three types of pion, having positive, negative, or zero charge. The charged pions are antiparticles of each other. *See also* elementary particles.

pitch The sensation that a sound produces in a listener as a result of its frequency (though other factors are involved). High-pitched notes are high-frequency vibrations and low-pitched notes are low-frequency vibrations.

pitot tube A double tube placed with one end in a stream of fluid, parallel to the flow, to measure fluid pressure. It has two openings. One is usually at the side, in the outer section of a double-walled tube, and measures static pressure, p. The other, in the inner section, faces the fluid flow and registers the total pressure (the sum of p and dynamic pressure). The other ends are connected to pressure measuring devices. Pitot tubes can be used to measure fluid velocity. *See* Bernoulli effect.

Planck constant Symbol: h A fundamental constant; the ratio of the energy (W) carried by a photon to its frequency (ν). A basic relationship in the quantum theory of radiation is $W = h\nu$. The value of h is $6.626\ 196 \times 10^{-34}$

J s. The Planck constant appears in many relationships in which some observable measurement is quantized (i.e. can take only specific separate values rather than any of a range of values).

Planck's formula See Planck's radiation law.

Planck's radiation law This major step towards modern physics was the first successful description of the wavelength distribution of black-body radiation. Earlier theories could adequately formulate the black-body radiation curves at certain wavelengths; none, however, could explain how the curve behaves over the whole wavelength range.

Planck's solution was to propose the quantization of energy in radiation into packets called quanta or photons; previously all radiators were viewed as emitting continuous streams of waves.

The theory led to an expression (*Planck's formula*) for the energy radiated per unit area per unit time from a black body at a given wavelength:
$$M_V = 2\pi hc^2\lambda^{-5}/[\exp(hc/\lambda kT) - 1]$$
where h is the Planck constant, c the speed of light, λ the wavelength, T the absolute temperature, and k the Boltzmann constant.

plane of polarization For historical reasons, this is defined as the plane containing incident and reflected rays in cases of polarization by reflection. It is therefore the plane containing the magnetic vector B, rather than the electric vector E.

plane polarization A type of polarization of electromagnetic radiation in which the vibrations take place entirely in one plane. It can be produced by reflection, or by transmission through a Nicol prism or through Polaroid. *See* polarization.

plano-concave lens A diverging lens with one plane face and one concave face. *See also* lens.

plano-convex lens A converging lens with one plane face and one convex face. See also lens.

plasma A mixture of free electrons and ions or atomic nuclei. Plasmas occur in thermonuclear reactions, as in the Sun. The glowing region of ions and electrons in a discharge tube is also a plasma. Sometimes plasmas are referred to as a fourth state of matter.

plasticity The tendency of a material to suffer a permanent deformation; i.e. not to return to its original dimensions after a deforming stress has been removed. An elastic material becomes plastic above its yield point. See elasticity.

platinum black A finely divided black form of platinum produced, as a coating, by evaporating platinum onto a surface in an inert atmosphere. Platinum-black coatings are used as pure absorbent electrode coatings in experiments on electric cells. They are also used, like carbon-black coatings, to improve the ability of a surface to absorb radiation.

p-n-p transistor See transistor.

point defect See defect.

point source A source of exactly spherical wave fronts. All sources can be considered to be point-like if viewed from a large enough distance. The stars provide an obvious example.
In a number of practical situations, sources that are effectively point sources are required. Normally a small hole in an otherwise opaque illuminated surface is used. Holes can be made as small as $100 \ \mu m$; the main problem is to arrange adequate illumination of the hole by the real source.

poise Symbol: P A unit of dynamic viscosity in the c.g.s. system. It is equal to $0.1 \ N \ s \ m^{-2}$.

Poiseuille's formula (Hagen's formula) An equation that describes the laminar flow of liquid through a pipe. The volume flow per second in a circular pipe of radius r and length l is given by:
$$\pi r^4 \Delta p / 8 \eta l$$
where Δp is the fluid pressure difference and η is its viscosity.

Poisson ratio Symbol: ν A measure of how a material changes shape when it is stretched. It is equal to the fractional change in cross-sectional area ($\Delta a / a$) divided by the fractional change in length ($\Delta l / l$).

polarimeter (saccharimeter) A device for measuring the angle of rotation of a plane-polarized beam caused by an optically active sample. Typically, light from a source is passed through a Nicol prism (the polarizer) to plane-polarize it, then through the sample. The amount of rotation is measured by a second prism (the analyser), which can be turned on an angular scale. The position of minimum transmission is observed. As this angle depends, among other things, on the concentration of an active solute, polarimeters are used to measure this.

polarization The restriction of the vibrations in a transverse wave so that the vibration occurs in a single plane. Electromagnetic radiation, for instance, is a transverse wave motion. It can be thought of as an oscillating electric field and an oscillating magnetic field, both at right angles to the direction of propagation and at right angles to each other. Usually, the electric vector is considered since it is the electric field that interacts with charged particles of matter and causes the effects. In 'normal' unpolarized radiation, the electric field oscillates in all possible directions perpendicular to the wave direction. On reflection or on transmission through certain substances (e.g. Polaroid) the field is confined to a single plane. The radiation is then said to be plane-polarized. Other types of polarization can occur under certain circumstances. Electromagnetic radiation of long wavelengths (e.g. radio waves) is normally produced polarized. Infrared, visible, ultraviolet, and shorter-wave-length radiation is usually unpolarized. *See* circular polarization, elliptical polarization. *See also* birefringent crystal, Nicol prism, optical activity, plane of polarization.

polarization, angle of *See* Brewster angle.

polarization, electrolytic The reduction of current in a voltaic cell, caused by the build-up of products of the chemical reaction. Commonly, the cause is the build-up of a layer of bubbles (e.g. of hydrogen). This reduces the effective area of the electrode, causing an increase in the cell's internal resistance. It can also produce a back e.m.f. Often a substance such as manganese dioxide (a depolarizer) is added to prevent hydrogen build-up.

polarization, electric The separation of the charges in the molecules of an insulator as an effect of an electric field. One face of an insulator in a field gains a net positive charge with the other becoming negative.

polarizer A device or arrangement producing plane-polarized radiation from incident unpolarized radiation. A Nicol prism or Polaroid sheets are normally used.

polarizing angle *See* Brewster angle.

Polaroid (*Trade name*) A synthetic doubly refracting substance, that strongly absorbs polarized light in one plane, while easily passing polarized light in another plane at right angles. Unpolarized light passed through a sheet of Polaroid is plane-polarized. Spectacle lenses made of this material normally absorb light vibrating horizontally — as produced by reflection from horizontal surfaces. They thus reduce reflected glare.

The first type was announced by Land in 1934. The modern form is made from a plastic sheet, highly strained to align the molecules and make it birefringent,

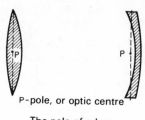

P-pole, or optic centre

The pole of a lens

then stained with iodine to make it dichroic.

pole (optical centre, optic centre) The geometric centre of a surface or a lens. In ray diagrams rays can be drawn undeviated through the pole of a lens. At the pole of a mirror the incident and reflected rays make equal angles with the principal axis. All distances from a refracting surface, reflector, or lens — object and image distances, focal distances, radii of curvature, etc. — are measured from the pole. *See also* lens, mirror.

poles, magnetic The regions of a magnetic field where the forces appear strongest. Thus, in a bar magnet, the lines of force appear to diverge from and converge to two small regions near the ends, these being the poles of the magnet. The concept of magnetic 'pole', however, has no more reality than that of 'regions in the field', as in the first sentence above.

A north-seeking pole (*north pole* or *N-pole*) is attracted approximately in the direction of geographical North; a south-seeking pole (*south pole* or *S-pole*) tends to move towards geographical South. Magnetic poles can only be found in opposite (unlike) pairs. The Earth's magnetic poles — regions where the Earth's field is strongest — are close to the geographical poles.

pole strength The force exerted by a given magnetic pole when it is one metre from a unit pole in a vacuum. A unit pole is defined as that pole which, when placed one metre from a similar pole in a vacuum, exerts a force of one newton.

polyatomic Describing a molecule that consists of several atoms (three or more). Examples are benzene (C_6H_6) and methane (CH_4).

polychromatic radiation Electromagnetic radiation that has a mixture of different wavelengths. *Compare* monochromatic radiation.

population inversion *See* laser.

positive *See* charge.

positive column A luminous area in an electrical discharge in a gas, occurring near the positive electrode.

positive feedback *See* feedback.

positive lens A lens with a positive power. *See* converging lens.

positive mirror A curved mirror with a positive power. *See* converging mirror.

positive pole A magnetic N-pole. *See* pole.

positive rays Streams of positive ions in an electrical discharge.

positron The antiparticle of the electron; a particle with the same mass as the electron but with a positive charge.
Positrons are found in cosmic-ray showers, where they result from pair production. They are also produced by a type of beta decay. They are annihilated when they encounter an electron, so have a short separate existence. Positrons can be detected by their tracks in cloud or bubble chambers, where they show the opposite deflection to electrons in a magnetic field.
See elementary particles.

positronium A fleeting combination of an electron and a positron to form an analogue to a hydrogen atom. When the two particles have their spins parallel the half-life is about 1.5×10^{-7} s; when they are antiparallel the half-life is shortened to 10^{-10} s. A positronium 'atom' decays to form two photons by annihilation. A combination of two electrons and two positrons also appears to exist, and is known as a positronium 'molecule', analogous to a hydrogen molecule.

Post Office box A box containing resistances that can be switched into the circuit, suitable for use as a Wheatstone bridge or potentiometer. *See* Wheatstone bridge.

potassium–argon dating A method of radioactive dating used for estimating the age of certain rocks. It is based on the decay of ^{40}K to ^{40}Ar (half-life 1.28×10^9 years). The technique is useful for time periods from 10^5 years ago back to the age of the Earth (4.6×10^9 years).

potential difference The difference in electrical potential between two points. *See also* electromotive force.

potential difference (p.d.) Symbol: V The difference in electric potential between two points in an electric field or in an electric circuit. The energy transferred, in joules, in taking Q coulombs of charge across a p.d. of V volts is $W = QV$, regardless of the path taken. *See also* electromotive force.

potential divider *See* voltage divider.

potential energy Symbol: V The work an object can do because of its position or state. There are many examples. The work an object at height can do in falling is its gravitational potential energy. The energy 'stored' in elastic or a spring under tension or compression is elastic potential energy. Potential difference in electricity is a similar concept, and so on. In practice the potential energy of a system is the energy involved in bringing it to its current

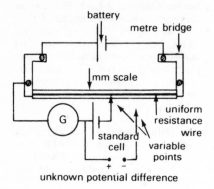

Potentiometer

state from some reference state, or vice versa.
See also energy.

potentiometer An instrument for measuring electrical potential differences by balancing two opposing potentials so that no current flows through a galvanometer. A standard reference cell with a known e.m.f. is connected across the ends of a uniform resistance wire. The unknown e.m.f. is then connected, so that it opposes the e.m.f. of the standard cell, between one end, A, of the wire and a sliding contact that can move along the wire. A galvanometer is connected in series with this cell.

When the position of the contact is adjusted to make the reading on the galvanometer zero, the length of wire from A to the contact, divided by the total length, is equal to the ratio of the unknown e.m.f. to that of the reference cell. This is a very accurate method because no current flows across the potential being measured.

Potentiometers are used to compare two e.m.f.s by balancing each in turn against the reference cell. They can also be used to measure current, by measuring the potential drop across a standard resistor. Another application is in comparing two resistances by measuring the potential difference across each in turn when

the same current flows through them. *See also* metre bridge.

pound An Imperial unit of mass. Formerly, it was defined as the mass of a platinum prototype. In 1963 the definition was changed to 0.453 592 37 kilogram.

poundal Symbol: pdl The unit of force in the f.p.s. system. It is equal to 0.138 255 N.

power (focal power) Symbol: P A measure of the ability of a lens or mirror to converge a parallel beam, given by the reciprocal of the focal distance f.

$$P = 1/f$$

Generally, f is in metres in which case P is in dioptres (D).

Strictly, power measures the ability of a reflecting or refracting surface, or of a lens, to change the curvature of incident wave fronts. The dioptre is better called the radian per metre (rad m^{-1}).
See also dioptre.

power Symbol: P The rate of energy transfer (or work done) by or to a system. The unit of power is the watt — the energy transfer in joules per second. In an electrical system power is given by VI, where V is the potential difference across a conductor and I the current through it. If V and I are not in phase the power absorbed is $IV\cos\phi$, where ϕ, the phase angle, is known as the 'power factor'.

Poynting vector Symbol: S The vector product of the electric field vector E and the magnetic field vector B in an electromagnetic wave. The Poynting vector gives, in magnitude and direction, the power radiated through unit area at any instant. The unit is the watt per square metre. The average value is give by $\frac{1}{2}E_0B_0$, E_0 and B_0 being the amplitudes of E and B.

precession If a body is spinning on an axis, the axis of rotation can itself move around another axis at an angle to it.

The effect is seen in gyroscopes. It also occurs for closed orbits (e.g. of a planet), in which the whole orbit moves around an axis.

presbyopia The normal development with age of far sightedness. The distance to the eye's near point increases as the eye's ability to accommodate is reduced (the lens becomes less elastic with age). See near point.

pressure Symbol: p The pressure on a surface due to forces from another surface or from a fluid is the force acting at 90° to unit area of the surface: pressure = force/area. The unit is the pascal (Pa).
Objects are often designed to maximize or minimize pressure applied. To give maximum pressure, a small contact area is needed — as with drawing pins and knives. To give minimum pressure, a large contact area is needed — as with skis and the large tyres of large vehicles. Where the pressure on a surface is caused by the particles of a fluid (liquid or gas), it is not always easy to find the force on unit area. The pressure at a depth in a fluid is the product of the depth, the average fluid density, and g (the acceleration of free fall).
Pressure in a fluid = depth × mean density × g
As it is normally possible to measure the mean density of a liquid only, this relation is usually restricted to liquids.
The pressure at a point at a certain depth in a fluid:
(1) is the same in all directions;
(2) applies force at 90° to any contact surface;
(3) does not depend on the shape of the container.
See also gas laws, pressure of the atmosphere, upthrust.

pressure broadening The increase in the width of a spectral line resulting from collisions between atoms or molecules. It relates to pressure, as there are more collisions at high pressure. See monochromatic radiation.

pressure of the atmosphere The pressure at a point near the Earth's surface due to the weight of air above that point. Its value varies around about 100 kPa (100 000 newtons per square metre). Barometers are used to measure atmospheric pressure which is important because the small changes relate to imminent weather changes.
Because the pressure at depth in a fluid depends on depth, barometers can be used as altimeters; they can be marked to indicate distance above or below sea level.

pressurized-water reactor (PWR) A nuclear reactor in which water, used as moderator, coolant, and reflector, is kept under pressure to prevent it from boiling at the operating temperature. The core is of 3% enriched uranium. This type of reactor has been used in nuclear powered submarines from 1954 onwards.

Prévost's theory of exchanges The theory that all bodies emit and absorb radiation continuously. The rate of emission depends on the body's surface and its temperature. The rate of absorption depends on the surface and the temperature of the surroundings. When the temperature of a body is constant, it is losing and gaining energy at equal rates, and is in equilibrium with its surroundings. If there is a temperature difference, there is a net flow of energy (heat).

primary cell A voltaic cell in which the chemical reaction that produces the e.m.f. is not reversible. This is not the case with a secondary cell (accumulator).

primary colours A set of three coloured lights (hues) that when mixed together in the right proportions, give the sensation of white. Normally the hues are chosen from the red, green, and blue regions.
An infinite number of sets of primary hues exists. The standard choice is one of convenience. According to the

primary winding

trichromatic theory of colour, any set of three primary colours has only one requirement — no one of them can be matched by mixing the other two.
See also colour.

primary winding *See* transformer (voltage).

principal axis (axis) The line joining the centres of curvature of the faces of a lens, or the line normal to a reflector at the pole. *See also* lens, mirror.

principal focus *See* focal point.

principal plane The plane perpendicular to the principal axis of a lens centred on the pole (optical centre). The pole is equidistant between the principal focal points. This is strictly true only for a thin lens. A thick lens has two principal planes; each is centred on a principal point. The focal distance of a thick lens is the distance between each focal point and the principal point on that side of the lens.

principal points Two points of a thick lens corresponding to the single pole (optical centre) of a thin lens. *See* principal plane.

principle of moments When an object or system is in equilibrium, the sum of the moments in any direction equals the sum of the moments in the opposite direction. Because there is no resultant turning force, the moments of the forces can be measured relative to any point in the system or outside it.

principle of superposition When two (or more) waves of the same type pass through the same region, the amplitude of vibration at any point is the algebraic sum of the individual amplitudes. Sign (crest or trough) must be taken into account. The waves emerge from the region of superposition unaffected. *See also* interference.

printed circuit A circuit that is constructed as an entity on an insulating board without the need for connecting wires, the connections being incorporated on the surface of the board as a conducting film. The board is first completely coated with a conductor film, usually copper. The circuit is then 'printed' on the board by a photographic process that superimposes another film, this time of protective material in a predetermined pattern. The exposed conducting film is then removed by etching, and the remainder acts as a network of conductors between components or ICs built into the circuit.
Boards are often printed on both sides, and some have two or more conducting layers, insulated from each other and separately etched. *See also* integrated circuit.

prism binoculars *See* binoculars.

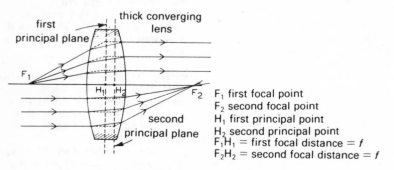

F₁ first focal point
F₂ second focal point
H₁ first principal point
H₂ second principal point
F₁H₁ = first focal distance = *f*
F₂H₂ = second focal distance = *f*

Principal plane

150

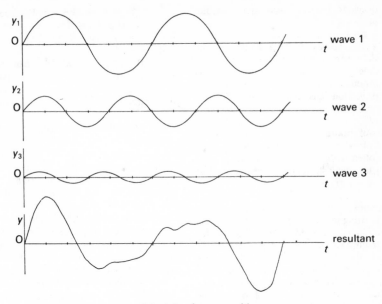

Principle of superposition

prism, optical An important component of many optical systems: a block (usually of glass for work with visible radiation) with two nonparallel sides. Prisms are often triangular in section. Prism action depends on two successive refractions, each with deviation (bending) in the same direction. This is effective, for instance, in spreading different wavelengths into a spectrum. Many prisms have been designed for special purposes for use in optical instruments. *See* binoculars, erecting prism. *See also* deviation.

progressive wave *See* wave.

projectile An object falling freely in a gravitational field, having been projected at a speed v and at an angle of elevation θ to the horizontal. In the special case that $\theta = 90°$, the motion is linear in the vertical direction. It may then be treated using the equations of motion. In all other cases the vertical and horizontal components of velocity must be treated separately. In the absence of friction, the horizontal component is constant and the vertical motion may be treated using the equations of motion. The path of the projectile is an arc of a parabola. Useful relations are:

time to reach maximum height
$$t = v\sin\theta/g$$
maximum height
$$h = v^2\sin^2\theta/2g$$
horizontal range
$$R = v^2\sin2\theta/g$$

Note that similar treatment can be applied to the case of an electric charge projected into an electric field. *See also* orbit.

projectors Optical arrangements able to produce a large bright real image on a screen.

There are two main types of projector — those in which light from the source is passed through the object, and those in which the light is reflected by the object. The *opaque projector* (episcope) is the main example of the latter. The former type, the diascope, includes the overhead projector and slide and film projectors.

Light from the source, concentrated by a converging mirror, passes through the slide (or is reflected by the opaque object). A condenser (condensing lens) produces an image; the converging projection lens forms an image of this on the screen.

proof plane A small shaped piece of foil with an insulating handle, used (often with an electroscope) to investigate the distribution of charge on the surface of an object.

proportional counter A counter for ionizing radiation in which the potential difference applied is high enough for multiplication of ions, so that the height of the pulse is proportional to the number of ions produced by the particle, and thus to its energy loss. A counter used in this way is said to be in the *proportional region*.

proportional limit *See* Hooke's law, elasticity.

proportional region *See* proportional counter.

proton An elementary particle with a positive charge (+ 1.602 192 C) and rest mass $1.672\,614 \times 10^{-27}$ kg. Protons are nucleons, found in all nuclides. *See also* elementary particles.

proton number (atomic number) Symbol: Z The number of protons in the nucleus of an atom. The proton number determines the chemical properties of the element because the electron structure, which determines chemical bonding, depends on the electrostatic attraction to the positively charged nucleus.

proton–proton chain reaction A series of nuclear fusion reactions by which hydrogen is converted into helium. It is thought that the reactions are responsible for energy production in the Sun and similar stars. There are in fact three possible sequences. The main one is:

$$^1H + {}^1H \rightarrow {}^2H + \nu + e^+$$
$$^2H + {}^1H \rightarrow {}^3He + \gamma$$
$$^3He + {}^3He \rightarrow {}^4He + 2\,{}^1H$$

the two other sequences lead to ^4He also, through ^7Li and ^8Be. *See also* carbon cycle.

psi particle *See* elementary particles.

p-type conductivity *See* semiconductor, transistor.

pulley A class of machine. In any pulley system power is transferred through the tension in a string wound over one or more wheels. The force ratio and distance ratio depend on the relative arrangement of strings and wheels. Efficiency is not usually very high as work must be done to overcome friction in the strings and the wheel bearings, and to lift any moving wheels. There are two types of single pulley system and two types of multiple pulley system. In the picture, T is string tension; in each case it is assumed that efficiency is 100%.

pulsatance *See* angular frequency.

pulse generator *See* astable circuit.

pupil The aperture of the eye, adjustable in size by the circular iris muscle. The pupil appears black as very little incident light is reflected by the retina.

pure note (pure tone) *See* note.

PWR *See* pressurized-water reactor.

pyrolysis The decomposition of chemical compounds by subjecting them to very high temperature.

pyrometer A device for measuring very high temperatures (above about 1000°C) by the radiation emitted by the body.
There are two main types. *See* optical pyrometer, total-radiation pyrometer.

pyrometry The measurement of high temperatures using pyrometers.

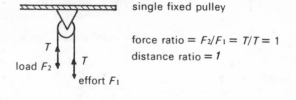

single fixed pulley

force ratio = $F_2/F_1 = T/T = 1$
distance ratio = 1

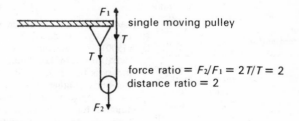

single moving pulley

force ratio = $F_2/F_1 = 2T/T = 2$
distance ratio = 2

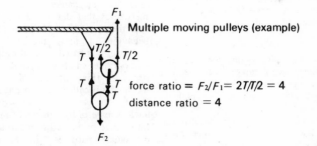

Multiple moving pulleys (example)

force ratio = $F_2/F_1 = 2T/T2 = 4$
distance ratio = **4**

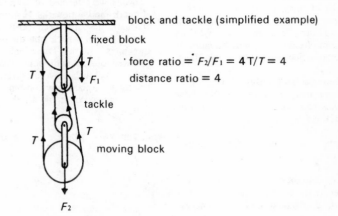

block and tackle (simplified example)

fixed block

force ratio = $F_2/F_1 = 4\,T/T = 4$
distance ratio = 4

tackle

moving block

Q

quality of sound (timbre) The characteristic of a musical note that is determined by the frequencies present. It enables a listener to tell the difference between two notes of the same fundamental frequency played on different musical instruments. A pure tone has no overtones present and its waveform is a sine wave. Musical instruments produce notes that have more complex waveforms, and it can be shown that these are formed by mixing a pure tone (the fundamental) with different higher frequencies. These frequencies are simple multiples of the fundamental frequency (f); i.e. $2f$, $3f$, and so on. The fundamental together with the 'overtones' are harmonics of the note; the different contributions characterize its quality.

quantized A physical quantity is said to be quantized when it can only change in definite steps — it does not have a continuous range of values. To explain the photoelectric effect, for instance, the energy of the electromagnetic radiation must be quantized. Angular momentum is quantized in atoms and molecules.

quantum (plural *quanta*) A definite amount of energy released or absorbed in a process. Energy often behaves as if it were 'quantized' in this way. The quantum of electromagnetic radiation is the photon. *See also* photoelectric effect.

quantum electrodynamics The use of quantum mechanics to describe how particles and electromagnetic radiation interact.

quantum mechanics *See* quantum theory.

quantum number An integer or half integer that specifies the value of a quantized physical quantity (energy,

angular momentum, etc.). *See* atom, Bohr theory, spin.

quantum states States of an atom, electron, particle, etc., specified by a unique set of quantum numbers. For example, the hydrogen atom in its ground state has an electron in the K shell specified by the four quantum numbers: $n = 1$, $l = 0$, $m = 0$, $m_s = 1/2$.

quantum theory A mathematical theory originally introduced by Max Planck (1900) to explain the black-body radiation from hot bodies. Quantum theory is based on the idea that energy (or certain other physical quantities) can be changed only in certain discrete amounts for a given system. Other early applications were the explanations of the photoelectric effect and the Compton effect and the Bohr theory of the atom. *Quantum mechanics* is a system of mechanics that developed from quantum theory and is used to explain the behaviour of atoms, molecules, etc. In one form it is based on de Broglie's idea that particles can have wavelike properties — this branch of quantum mechanics is called *wave mechanics*.

quark A type of hypothetical fundamental particle postulated to make up other elementary particles. In the original theory there were three types having charges -1/3 or +2/3 the proton charge. Hadrons were formed of combinations of quarks and antiquarks. A fourth type of quark has also been postulated to explain the psi particle and the property known as 'charm'. Theoreticians have endowed quarks with other properties such as 'colour', 'flavour', 'truth', and 'beauty' to explain how combinations of quarks can account for all the known hadrons.

quarter-wave plate *See* retardation plate.

R

rad A unit of absorbed dose of ionizing radiation, defined as being equivalent to an absorption of 10^{-2} joule of energy in one kilogram of material.

radar An acronym for *radio direction* and *ranging*. A system for the precise location of distant objects (especially ships and aircraft) by means of radio. A microwave radio beam is transmitted, reflected by the distant object, and received on return to the transmitter. The directive aerial points in the direction of the object, the distance of which is calculated from the time of travel of the wave at the speed of electromagnetic waves, 3×10^8 m s^{-1}. Alternatively the beam can be made to scan in continuous circles, and so locate all objects in any direction, the locations being plotted on a plan position indicator (PPI), a cathode ray tube with a rotating time base. *See also* radio.

radian Symbol: rad The SI unit of plane angle; 2π radian is one complete revolution (360°).

radiance Symbol: L_e In a given direction, the radiant intensity of a source of radiation per unit area projected at right angles to the direction. The unit is the watt per steradian per square metre (W sr^{-1} m^{-2}).

radiant exitance *See* exitance.

radiant flux Symbol: Φ_e The rate of flow of radiant energy; the total power emitted or received in the form of electromagnetic radiation. It is measured in watts.

radiant heat Energy transmitted as infrared radiation emitted from a body at normal temperatures. *See also* black body.

radiant intensity Symbol: I_e The radiant flux from a point source per unit solid angle. The unit is the watt per steradian (W sr^{-1}).

radiation In general the emission of energy from a source, either as waves (light, sound, etc.) or as moving particles (beta rays or alpha rays). The term is used in two restricted ways:
1. The transfer of energy as electromagnetic radiation; traditionally said to be one of the three ways in which heat transfer occurs (the others quoted are conduction and convection). *See also* infrared radiation.
2. Particles (alpha or beta particles) or photons (gamma rays) emitted from a radioactive material.

radiation formula *See* Planck's radiation law.

radiation pressure Pressure exerted on a surface by electromagnetic radiation. Since the radiation carries momentum as well as energy it exerts a force when it meets a surface. For example, the radiation pressure of sunlight on the Earth is about 10^{-5} Pa.

radio The process of communication across space by the transmission and reception of an electromagnetic wave of radiofrequency without the use of connecting wires or other material link.

radioactive Describing an element or nuclide that exhibits natural radioactivity.

radioactive dating (radiometric dating) Any method of measuring the age of materials that depends on radioactivity. *See* carbon dating, dating, fission-track dating, potassium–argon dating, rubidium–strontium dating, thermoluminescent dating, uranium–lead dating.

radioactive series A series of radioactive nuclides, each being formed by the decay of the previous one. All such series terminate in the product of a stable nuclide. An example of a radioactive series is the *thorium series*, which starts with ^{232}Th, and which, after six alpha decays and four beta decays, terminates in ^{208}Pb. This series is known as the 4*n* series because all its members have mass numbers that are multiples of 4. The *uranium series*, passing from ^{235}U to ^{207}Pb is known as the 4*n* - 1 series. The *actinium series* and the *neptunium series* are two other radioactive series.

radioactivity The disintegration of certain unstable nuclides with emission of radiation. There are various types classified as alpha, beta, and gamma decay, and fission.

radiocarbon dating *See* carbon dating.

radio frequency (RF) A frequency between 3 KHz and 300 GHz (wavelength 1 cm-100 km).

radioisotope A radioactive isotope of an element. Tritium, for instance, is a radioisotope of hydrogen. Radioisotopes are extensively used in research as sources of radiation and as tracers in studies of chemical reactions. Thus, if an atom in a compound is replaced by a radioactive nuclide of the element (a *label*) it is possible to follow the course of the chemical reaction. Radioisotopes are also used in medicine for diagnosis and treatment.

radioluminescence *See* luminescence.

radiometric dating *See* radioactive dating.

radio waves A form of electromagnetic radiation with wavelengths greater than a few millimetres. There is no upper limit — applications have been found for radio waves tens of kilometres long.
Artificial radio waves of any frequency can be produced when suitable electronic circuits feed alternating signals to suitable aerials. They are generated by the oscillating motion of the electrons in the conductor.
The main application of radio waves is to carry information — for 'wireless telegraphy' and television for instance.
  The information is used to modulate a carrier wave, suitably changing its amplitude, frequency, or phase during transmission. The receiver is able to separate the carried information from the carrier.
See also electromagnetic spectrum.

radius of curvature Symbol: *r* The radius of the sphere of which a mirror or lens surface is part. In the case of a reflector, the radius of curvature is twice the focal distance, *f*.
$$r = 2f$$
In the case of a lens it is of value to use the points at 2f and 2f', twice as far from the pole as f and f'. The distance to either of these, double the focal distance, is the nominal radius of curvature of the lens. It can be related to the radii of curvature of the two lens surfaces.
See also lens, mirror.

radius of gyration Symbol: *k* For a body of mass *m* and moment of inertia *I* about an axis, the radius of gyration

Frequency bands of radio waves

Band		Frequency range (Hz)	Wavelength range (m)
extremely low frequency	(ELF)	$- 3 \times 10^3$	$10^5 -$
very low frequency	(VLF)	$3 \times 10^3 - 3 \times 10^4$	$10^4 - 10^5$
low frequency	(LF)	$3 \times 10^4 - 3 \times 10^5$	$10^3 - 10^4$
medium frequency	(MF)	$3 \times 10^5 - 3 \times 10^6$	$10^2 - 10^3$
high frequency	(HF)	$3 \times 10^6 - 3 \times 10^7$	$10 - 10^2$
very high frequency	(VHF)	$3 \times 10^7 - 3 \times 10^8$	$1 - 10$
ultra high frequency	(UHF)	$3 \times 10^8 - 3 \times 10^9$	$0.1 - 1$

about that axis is given by $k^2 = I/m$. In other words, a point mass m rotating at a distance k from the axis would have the same moment of inertia as the body.

rainbow The effect of refraction in, for example, rain drops causing sunlight to separate into the colours of the spectrum. The Sun is behind the observer as he faces the rainbow.

In good viewing conditions more than the ordinary primary bow can be seen. The sky is noticeably darker above the primary bow; above this again is a much fainter secondary bow, caused by two internal reflections in each drop rather than one. Sometimes other bows (supernumerary bows) can also be seen.

ratio arms See Wheatstone bridge.

rationalized units A system of units in which the equations have a logical form related to the shape of the system. SI units form a rationalized system of units. In it formulae concerned with circular symmetry contain a factor of 2π; those concerned with radial symmetry contain a factor of 4π.

ray A very narrow beam of radiation. By considering how selected rays behave, ray diagrams can be drawn to relate object and image positions for any lens or mirror.

Theoretically a ray is infinitely narrow, like a mathematical line. Then, even if part of a convergent or divergent beam, it will not converge or diverge itself. There is no evidence that beams are really formed of rays in this way; however rays are extremely useful in discussion and in drawings.

ray diagrams Accurate or approximate diagrams showing how selected rays pass through an optical system (such as a lens, prism, or microscope). The rays are selected on the basis of known behaviour. Suitable ray diagrams can show details of the image formed of any object by the optical system. They are also important in system design. Many examples appear in this book.

Rayleigh criterion See resolving power.

reactance Symbol: X A measure of the opposition of inductance or capacitance to alternating current. Reactance is the ratio of the peak voltage to the peak current. Like resistance it is measured in ohms; it causes the current to become out of phase with the voltage. Reactance depends on the frequency of the supply. For a pure capacitance (C), the current leads the voltage by one quarter cycle. The reactance is given by $1/2\pi fC$, where f is the frequency. For a pure inductance the current lags behind the voltage by one quarter cycle. The reactance is given by $2\pi fL$. See also alternating-current circuit, impedance, resonance.

reaction Newton's third law of force states that whenever object A applies a force on object B, B applies the same force on A. An old word for force is 'action'; 'reaction' is thus the other member of the pair.

Often it is hard, or impossible, to decide what A and B are. Thus in the interaction between two electric charges, each exerts a force on the other. Thus, in general, action and reaction have little meaning. The word reaction is still sometimes used in restricted cases, such as the reaction of a support on the object it supports. In this case the action is the effect of the weight of the object on the support.

reactor See nuclear reactor.

real image See image.

real is positive, virtual is negative See sign convention.

rectifier An electrical conductor that allows current to flow through it in one direction only, thus enabling the conversion of a.c. to d.c. The commonest type of rectifier is the semiconductor diode, which consists of a single p-n junction. A metal rectifier, which is also a junction rectifier, consists of a metal in contact with a semiconductor or an oxide of the metal. Typical examples are copper-

selenium and copper-copper oxide recti-
fiers. The thermionic diode is another
type of rectifier.

rectilinear propagation Passage of
radiation in straight lines in a constant
medium.

red shift A displacement of lines in the
spectra of certain celestial objects
towards longer wavelengths (i.e. towards
the red end of the visible spectrum). The
spectral lines appear at slightly longer
wavelengths than they would under
'normal' laboratory conditions. It is
caused by the Doppler effect and indicates that the observed galaxy is moving
away from the Earth. Some objects
show a *blue shift*, indicating movement
towards the observer. These shifts are
known as *Doppler shifts*.

Not all red shifts are thought to be
caused by the Doppler effect. Some may
result from a high gravitational field.
These are explained by the General
theory of relativity, and are known as
gravitational red shifts or *Einstein shifts*.

reflectance (reflection factor) Symbol: ρ
The fraction of incident radiation
reflected by a surface; i.e. the ratio of
the reflected flux to the incident flux.
For each surface this will depend on
wavelength; often too it depends on
angle of incidence. There is no unit. If
the material is sufficiently thick, so that
the reflecting properties do not depend
on thickness, it is called *reflectivity*.

reflection The process in which radiation
meeting the boundary between two
media 'bounces back', to stay in the first
medium. Any kind of radiation — wave
or stream of particles — can be reflected.
Whenever an incident ray is reflected,
the laws of reflection can be used to find
the direction into which the ray is turned
(deviated). Reflection from a source
makes the radiation appear to come
from somewhere else — the image of
the source. For reflection by a plane
surface:
(1) The image is the same distance
behind the surface as the object is in
front.
(2) The line joining each object point
with its image is normal (perpendicular)
to the surface.
(3) The image is the same size as the
object.
(4) The image is the same way up as the
object, but perverted (laterally inverted).
(5) The image is virtual (no light actually
passes through it).

For reflection by a curved surface, the
relations between object and image
depend on how far the object is from the
surface compared with the focal distance
(f) of the surface. According to the real
is positive, virtual is negative convention, the image distance (v) and object
distance (u) are related by:
$$1/f = 1/v + 1/u$$
Image size (y) relates to object size (x)
thus:
$$y = x(v/u)$$

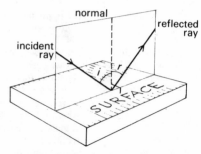

Angle of reflection

The focal distance of a curved reflector is half the radius of curvature (*r*).

Regular reflection (*specular reflection*) occurs from smooth surfaces. The beam is not scattered on reflection and undistorted images are formed. In *diffuse reflection* the reflecting surface is rough and the reflected beam is scattered randomly.

See also mirror, total internal reflection.

reflection, angle of *See* angle of reflection.

reflection factor *See* reflectance.

reflection, laws of Two laws that apply whenever any form of radiation is reflected, whatever the circumstances. In order to state the laws, the *normal* is defined — the perpendicular to the surface at the point of incidence. All angles named are measured from the normal.

(1) The reflected ray is in the same plane as the incident ray and the normal.

(2) The angle of reflection equals the angle of incidence and lies on the other side of the normal.

reflection, total internal *See* total internal reflection.

reflectivity *See* reflectance.

reflector *See* mirror.

reflector (reflecting telescope) An optical telescope in which the objective (the light collector) is a converging reflector. The largest ones in use have mirrors several metres in diameter. The problem with reflecting telescopes is that the first image is formed in the incident beam. There are a number of types distinguished by different optical systems for avoiding this and taking the light to the eyepiece. *See* Cassegrainian telescope, Coudé telescope, Gregorian telescope, Newtonian telescope.

refracting telescope *See* refractor.

refraction The process in which radiation incident on the boundary between two media passes on into the second medium. Any kind of radiation — wave or stream of particles — can be refracted. Unless the radiation meets the surface at a right angle (down the normal) it will change direction on refraction. The behaviour of refraction in any case depends on the refractive constant (index) of the pair of media. Whenever a ray is refracted, the laws of refraction can be used to find the direction into which the ray is turned (deviated).

Diffuse refraction occurs when the radiation is scattered on transmission through the medium, as on passage through translucent media such as wax or frosted glass. The opposite effect is *regular refraction*, in which the beam is not scattered and undistorted images can be formed.

See also refractive constant.

refraction, angle of *See* angle of refraction.

refraction, laws of Two laws that apply whenever any radiation is refracted, whatever the circumstances. In order to state the laws, the normal is defined — the perpendicular to the surface at the point of refraction. All angles named are measured from the normal.

(1) The refracted ray is in the same plane as the incident ray and the normal.

(2) The sine of the angle of incidence divided by the sine of the angle of refraction is constant for a given wavelength and a given pair of media; the angles lie on opposite sides of the normal.

The second law is sometimes called *Snell's law*. It leads to a definition for the refractive constant (*n*) concerned with a given wavelength and a given pair of media.

$$n = (\sin i)/(\sin r)$$

Special care needs to be taken with the laws of refraction in situations related to polarization of the radiation.

See also refractive constant, birefringent crystal.

refractive constant (refractive index)

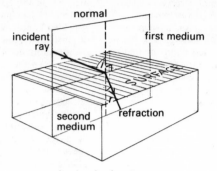

Angle of refraction

Symbol: n When radiation travels from one medium to another, refraction occurs. The refractive constant for the two media is the ratio of the speed of radiation in the first medium to that in the second:

$$_1n_2 = c_1/c_2.$$

Here, $_1n_2$ is the refractive constant for radiation passing from medium 1 to medium 2. Strictly, it depends on the wavelength and is often defined for yellow light (sodium D-lines).

The *absolute refractive constant* of a given material is that for the electromagnetic radiation passing from free space into the material:

$$n = c_0/c$$

where c_0 is the speed of the radiation in a vacuum and c the speed in the material. For successive refractions at a series of boundaries between media 1, 2, and 3 for instance:

$$_1n_3 = {}_1n_2 \times {}_2n_3$$

Other relations arise in certain circumstances. When the radiation changes direction during refraction:

$$n = (\sin i)/(\sin r)$$

where i is the angle of incidence and r the angle of refraction.

The apparent depth of a transparent medium is related to its real depth by:

$$n = \text{real depth/apparent depth}$$

When total internal reflection occurs at a critical angle c:

$$n = 1/\sin c$$

refractive index *See* refractive constant.

refractivity A measure of the ability of a medium to bend (deviate) radiation entering its surface.

refractometer A device for measuring the refractive constant of a medium. There are many types, each using one or other of the relations giving the refractive constant, directly or as applied in special circumstances. In the elementary laboratory, methods based on the apparent-depth and critical-angle expressions are most effective.

refractor (refracting telescope) An optical telescope in which the objective (the light collector) is a converging lens. Practical problems preclude the design of such telescopes with lenses more than about one metre in diameter. Telescopes used in astronomy produce an inverted image; Keplerian telescopes are invariably used. For non-astronomical use a *terrestrial telescope* is employed — either a Galilean telescope or a Keplerian telescope with an added inverting lens. *See* telescope, Galilean telescope, Keplerian telescope.

refrigerant The working fluid in a refrigerator.

refrigerator A device for producing a low temperature. Refrigerators use a volatile liquid, which vaporizes and expands through a narrow opening. The cooling occurs by the latent heat of vaporization of the liquid. The vapour is condensed back to liquid by compression with a pump. The process occurs in a continuous cycle.

Regnault's apparatus An apparatus for measuring the absolute thermal expansivity of a liquid. It consists of a shaped U-tube (see illustration) with one limb at ice temperature and the other at a constant temperature θ (usually the steam temperature). The value of γ can be calculated from the difference between the liquid heights in the two limbs of the tubes.

regular reflection *See* reflection.

regular refraction *See* refraction.

rejector circuit *See* resonance.

relative atomic mass Symbol: A_r The ratio of the average mass per atom of the naturally occurring element to 1/12 of the mass of an atom of nuclide ^{12}C. It was formerly called *atomic weight.*

relative density Symbol: *d* The density

of a substance divided by the density of water. Usually the density of water at 4°C (the temperature at which its density is a maximum) is used. The temperature of the substance is stated or is understood to be 20°C. Relative density was formerly called *specific gravity.*

relative humidity A measure of the amount of water vapour in the air, expressed as a proportion of the maximum amount the air could hold at the temperature concerned. (The higher the temperature the more water vapour air can hold.) For a fixed amount of water and air, the relative humidity therefore decreases as temperature rises and increases as it falls. If the temperature falls to the point where the relative humidity equals one, the water vapour starts to condense out as droplets on surfaces or in the air (dew or mist). This temperature is the saturation temperature or the *dew temperature* (dew point). *Compare* absolute humidity.

relative molecular mass Symbol: M_r The ratio of the average mass per molecule of the naturally occurring form of an element or compound to 1/12 of the mass of an atom of nuclide ^{12}C. This was formerly called *molecular weight.* It

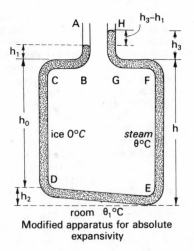

Modified apparatus for absolute expansivity

does not have to be used only for compounds that have discrete molecules; for ionic compounds (e.g. NaCl) and giant-molecular structures (e.g. BN) the formula unit is used.

relative permeability See permeability.

relative permittivity Symbol: ϵ_r For a parallel-plate capacitor (say) with a given material between the plates, the relative permittivity is the ratio of the capacitance observed (C_m) to the capacitance (C_0) if the material were removed (free space between the plates); i.e. $\epsilon_r = C_M/C_0$. Relative permittivity measures the effect of a material on an electric field relative to free space. It is equal to ϵ/ϵ_0, where ϵ is the (absolute) permittivity of the material and ϵ_0 the permittivity of free space (the electric constant). An early term for relative permittivity was *dielectric constant*. See also permittivity.

relative velocity If two objects are moving at velocities v_A and v_B in a given direction, the velocity of A relative to B is $v_A - v_B$ in that direction. In general, if two objects are moving in the same frame at non-relativistic speeds their relative velocity is the vector difference of the two velocities.

relativistic mass The mass of an object as measured by an observer at rest in a frame of reference in which the object is moving with a velocity v. It is given by $m_0 = m\sqrt{(1 - v^2/c^2)}$, where m_0 is the rest mass, c is the velocity of light, and m is the relativistic mass. The equation is a consequence of the Special theory of relativity, and is in excellent agreement with experiment. No object can travel at the speed of light because its mass would then be infinite. See also relativity, rest mass.

relativistic particle A particle that is travelling at a relativistic velocity. See relativistic velocity.

relativistic velocity Any velocity that is sufficiently high to make the mass of an object significantly greater than its rest mass. It is usually expressed as a fraction of c, the velocity of light. At a velocity of $c/2$ the relativistic mass of an object increases by 15% over the rest mass. See also relativistic mass, rest mass.

relativity, theory of A theory put forward in two parts by Albert Einstein. The Special theory (1905) referred only to non-accelerated (inertial) frames of reference. The General theory (1915) is also applicable to accelerated systems.

The *Special theory* was based on two postulates:

(1) That physical laws are the same in all inertial frames of reference.

(2) That the speed of light in a vacuum is constant for all observers, regardless of the motion of the source or observer.

The second postulate seems contrary to 'common sense' ideas of motion. Einstein was led to the theory by considering the problem of the 'ether' and the relation between electric and magnetic fields in relative motion. The theory accounts for the negative result of the Michelson–Morley experiment and shows that the Lorentz-Fitzgerald contraction is only an apparent effect of motion of an object relative to an observer, not a 'real' contraction. It leads to the result that the mass of an object moving at a velocity v relative to an observer is given by:

$$m = m_0/\sqrt{(1 - v^2/c^2)}$$

where c is the speed of light and m_0 the mass of the object when at rest relative to the observer. The increase in mass is significant at high velocities. Another consequence of the theory is that an object has an energy content by virtue of its mass, and similarly that energy has inertia. Mass and energy are related by the famous equation $E = mc^2$.

The *General theory* of relativity seeks to explain the difference between accelerated and non-accelerated systems and the nature of the forces acting in both of them. For example, a passenger in a car rounding a bend at constant speed is in an accelerated frame of reference because his velocity is changing. He experiences a fictitious force known as a centrifugal force. This force is fictitious because to an outside observer the force is simply a result of the passenger's tendency to continue in a straight line. This analysis of forces led Einstein to the view that gravitation is a consequence of the geometry of the universe. He visualized a four-dimensional space-time continuum in which the presence of a mass affects the geometry — the space is 'curved' by the mass.

reluctance The magnetic analogue of electric resistance. See magnetic circuit.

reluctance ('magnetic resistance') Symbol: R The ratio of magnetomotive force to total magnetic flux in a magnetic circuit. The SI unit of reluctance is henry metre^{-1} (H m^{-1}).

reluctivity The reciprocal of magnetic permeability.

rem (radiation equivalent man) A unit for measuring the effects of radiation dose on the human body. One rem is equivalent to an average adult male absorbing one rad of radiation. The biological effects depend on the type of radiation as well as the energy deposited per kilogram. The statutory limit for occupational exposure to radiation is 5 rem per year.

remanence See hysteresis cycle.

resistance box A device for providing variable known resistances. One type has a thick brass bar with gaps at intervals across which are connected fixed resistances. Combinations of these resistances can be chosen by inserting brass plugs in the gaps, thus shorting out selected resistors. Other types of resistance box have multi-switches to give the selection.

resistance, electrical Symbol: R The ratio of the potential difference across an electrical element to the current in it. Resistance measures the opposition of the component to the flow of charge.

In an alternating-current circuit resistance is only one factor in the response of the component. See impedance.
See also Ohm's law, effective resistance.

resistance thermometer A thermometer that measures temperature by the change in electrical resistance of a conductor. Usually a small coil of pure fine platinum wire is used, wound on a mica former. It is incorporated into one arm of a Wheatstone bridge, the other arm has a variable resistor and a pair of 'dummy' leads, to compensate for temperature effects in the leads of the measuring coil.
The resistance varies with temperature according to an approximate equation of the form:
$$R = R_0(1 + a\theta)$$
where R is the resistance at θ°C, R_0 the resistance at 0°C, and a is a constant.
More strictly:
$$R = R_0(1 + a\theta + b\theta^2)$$
where b is a second constant.

resistance wire Wire used to introduce a resistance into an electrical circuit. Resistance wire is made from alloys such as Nichrome.

resistivity Symbol: ρ The tendency of a material to oppose the flow of an electric current. The resistivity is a property of the material at a given temperature and does not depend on the sample. The resistance R of a sample is given by $\rho l/A$, where A is the cross section area and l the length. The unit is the ohm metre (Ωm). Resistivity was formerly called specific resistance.
Resistivities of conducting materials vary between about 10^{-8} and 10^{-6} Ωm. Semiconductors have values around 10^{-6}–10^{-7} Ωm and insulators values of 10^{-7} Ωm upwards.

resistor A component included in an electric circuit because of its resistance. Resistors may be variable or have a fixed value; they are made of resistance wire or carbon. Some types have a ceramic coating incorporating a colour

resolution

coding in the form of stripes, which identify the value of the resistance.

resolution The ability to distinguish two close objects separately, rather than as one single object. The resolution obtained with an optical system depends on the magnification and on the resolving power. *See* resolving power.

resolution of vectors The determination of the components of a vector in two given directions at 90°. The term is sometimes used in relation to finding any pair of components (not necessarily at 90° to each other).

resolving power A measure of the ability of an optical system to form detectably separate images of close objects or to separate close wavelengths of radiation.

The resolving power depends on the aperture (a) and the wavelength (λ) of the radiation. For separation of images it is the smallest angular separation of the images. It can be defined ideally as λ/a (the unit being the radian). In practice, diffraction effects limit the resolution. In a telescope forming images of two stars, each image will be a bright central portion surrounded by light and dark rings. The *Rayleigh criterion* for resolution is that the central portion of one image falls on the first minimum of the other. The resolving power is then $1.22\lambda/a$. The resolving power of a microscope is often given as the actual minimum distance between two points that can be separated (a small distance is a high resolving power). Again it depends on wavelength and aperture. Larger apertures give increased resolving power. The immersion objective is a method of increasing the effective aperture. Electron microscopes have higher resolving powers because the wavelengths of electrons are much smaller than those of visible radiation. The resolving power of the human eye is about 10^{-3} rad. At the standard near-point distance (250 mm) points about 100 μm apart can just be resolved.

resonance 1. The large-amplitude vibration of an object or system when given impulses at its natural frequency. For instance, a pendulum swings with a natural frequency that depends on its length and mass. If it is given a periodic 'push' at this frequency — for example, at each maximum of a complete oscillation — the amplitude is increased with little effort. Much more effort would be required to produce a swing of the same amplitude at a different frequency.

2. An electric circuit containing both capacitance and inductance can be arranged so that a very large (resonant) response occurs at a particular (resonant) frequency of alternating supply. An inductance (L) and capacitance (C) in series are known as a series resonant or *acceptor circuit,* and in it there is a large alternating current at the resonant frequency (f). Resonance occurs when $f.L = 1/f.C$ In a parallel resonant or *rejector circuit,* L and C are in parallel; then a minimum current flows at the resonant frequency.

In a tuning circuit, the resonant frequency may be varied by altering the capacitance or the inductance. This is how radio receivers are tuned to the broadcast frequency. *See also* impedance, reactance.

3. One of a large number of excited states of elementary particles. Many resonances are known; all have very short lifetimes.

Restitution constants

Type of collision	restitution constant (e)
perfectly inelastic	0
inelastic	>0, <1
perfectly elastic	1
superelastic	>1

restitution, coefficient of Symbol: e For the impact of two bodies, the elasticity of the collision is measured by the coefficient of restitution. It is the relative velocity after collision divided by the relative velocity before collision (velocities measured along the line of centres). For spheres A and B:

$$v_A^1 - v_B^1 = -e(v_A - v_B)$$

v indicates velocity before collision; v^1 velocity after collision. Kinetic energy is conserved only in a perfectly elastic collision. Different possibilities are shown in the table.

rest mass Symbol: m_0 The mass of an object at rest as measured by an observer at rest in the same frame of reference. *See also* relativistic mass.

resultant A vector with the same effect as a number of vectors. Thus, the resultant of a set of forces is a force that has the same effect; it is equal in magnitude and opposite in direction to the equilibrant. Depending on the circumstances, the resultant of a set of vectors can be found by different methods. *See* parallel forces, parallelogram of vectors, principle of moments.

retardation plate A thin transparent plate cut from a piece of birefringent material (e.g. quartz or mica) parallel to the optic axis. For light transmitted by the plate normal to the optic axis, the ordinary and extraordinary rays travel at different speeds. The thickness is designed to induce a desired phase difference between the transmitted rays. The *half-wave plate* introduces a phase difference of π, and the ordinary and extraordinary rays are out of step by one-half wavelength. The *quarter-wave plate* produces a phase difference of $\pi/2$ and the rays are out of step by one-quarter wavelength.

retina The light-sensitive inner surface of the eye: its function is to convert incident light into nerve signals. *See also* eye.

reverberation Persistence of sound after the source has stopped. Reverberation characteristics are important in the design of concert halls, theatres, etc.

reversible change A change in the pressure, volume, or other properties of a system, in which the system remains at equilibrium throughout the change. Such processes could be reversed; i.e. returned to the original starting position through the same series of stages. They are never realized in practice. An isothermal reversible compression of a gas, for example, would have to be carried out infinitely slowly and involve no friction, etc. Ideal energy transfer would have to take place between the gas and the surroundings to maintain a constant temperature.

In practice, all real processes are *irreversible changes* in which there is not an equilibrium throughout the change. In an irreversible change, the system can still be returned to its original state, but not through the same series of stages. For a closed system, there is an entropy increase involved in an irreversible change.

Reynolds number Symbol: *Re* A dimensionless quantity used to describe fluid flow. It is equal to $\rho c l/\eta$, where ρ is the fluid's density, c its velocity, η its viscosity, and l is a length characteristic of the geometry. For a long pipe, for example, l would be the internal diameter. The Reynolds number is used in making scale models of fluid flow. The size (l) may be scaled down by a factor of four, for instance, if the model fluid has a value of ρ/η equal to a quarter that of the fluid being modelled. Because *Re* is unchanged the flow pattern will be similar and the critical speed will be the same. *See also* critical speed.

RF *See* radio frequency.

rheology The study of fluid flow.

rheostat A variable resistor, usually operated by a sliding contact on a coil. *See also* potentiometer.

ribbon microphone *See* microphone.

right-hand rule *See* Fleming's rules.

rigidity modulus *See* elastic modulus.

ripple A device for producing waves (ripples) on the surface of water and viewing these under stroboscopic light. It is used to demonstrate interference effects and other wave properties.

RMS value *See* root-mean-square value.

Rochon prism A type of polarizing prism. The ordinary ray passes through undeviated; the extraordinary ray leaves from the side. This result is the reverse of that given by the Nicol prism.

rods The more sensitive cells of the retina of the eye. They are of most importance in vision in poor light, but cannot provide colour information. Their action is not known in detail. It appears to be based on the photochemical breakdown of a reddish dye called rhodopsin (formerly visual purple). *See also* eye.

roentgen Symbol: R A unit of radiation, used for X-rays and γ-rays, defined in terms of the ionizing effect on air. One roentgen induces 2.58×10^{-4} coulomb of charge in one kilogram of dry air.

Roentgen rays A former name for *X-rays*.

rolling friction *See* friction.

root-mean-square value (RMS value)
1. The square root of the average of the squares of a group of numbers or values. For example, the RMS value of $\{2, 4, 3, 2.5\}$ is $\sqrt{[(2^2 + 4^2 + 3^2 + 2.5^2)/4]}$
2. For continuous quantities, such as alternating electric current, the root-mean-square value is used to measure the average effect. The root-mean-square value of an alternating current (also called the *effective value*) is the value of the direct current that would produce the same energy transfer per second in a given resistor. For a sine wave, this is $I_m/\sqrt{2}$, where I_m is the peak value.

rotational motion Motion of a body turning about an axis. The physical quantities and laws used to describe

Rotational equations of motion

Variable	Equation
$a, \omega_1, \omega_2, t, (\theta)$	$\omega_2 = \omega_1 + at$
$\omega_1, \omega_2, t, \theta, (a)$	$\theta = (\omega_1 + \omega_2)/2t$
$a, \omega_1, t, \theta, (\omega_1)$	$\theta = \omega_1 t + at^2/2$
$a, \omega_2, t, \theta, (\omega_1)$	$\theta = \omega_2 t - at^2/2$
$a, \omega_1, \omega_2, \theta, (t)$	$\omega_2^2 = \omega_1^2 + 2a\theta$

linear motion all have rotational analogues; the equations of rotational motion are the analogues of the equations of motion (linear).
They normally include $T = I\alpha$, the analogue of $F = ma$, as well as the 'kinematic' equations themselves. Here T is the turning-force, or torque, I is the moment of inertia, and α is the angular acceleration.
The other equations relate the angular velocity, ω_1, of the object at the start of timing to its angular velocity, ω_2, at some later time, t, and thus to the angular displacement ϕ. *See* equations of motion.

rotor The rotating part, or armature, of an electric motor or dynamo, or similar device. *See also* stator.

rubidium–strontium dating A method of radioactive dating used for estimating the age of certain rocks. It is based on the decay of ^{87}Rb to ^{87}Sr (half-life 5×10^{10} years). The technique is useful for time periods from 10^7 years ago back to the age of the Earth (about 4.6×10^9 years).

Rydberg constant *See* Bohr theory.

Linear measures and their rotational analogues

Linear measure		Rotational analogue	
Quantity	Unit	Quantity	Unit
mass (m)	kg	moment of inertia (I)	kg m^2
displacement (s)	m	angular displacement (θ)	rad
velocity (v)	m s^{-1}	angular velocity (ω)	rad s^{-1}
acceleration (a)	m s^{-2}	angular acceleration (α)	rad s^{-2}
momentum (p)	kg m s^{-1}	angular momentum (L)	kg m^2 s^{-1}

S

saccharimeter *See* polarimeter.

saturated vapour pressure *See* vapour pressure.

saturation The freedom from white of a colour. A pure spectral colour (single wavelength) is a hue, and is said to be *saturated*. Unsaturated colours are tints; i.e. hues mixed with white.

saturation, magnetic A magnetic material is saturated if it can be magnetized no more strongly. This can easily be explained using the domain theory. *See also* hysteresis cycle.

scalar A measure in which direction is unimportant or meaningless. For instance, distance is a scalar quantity, whereas displacement is a vector. Mass, temperature, and time are scalars — they are each quoted as a pure number with a unit. *See also* vector.

scalar product A multiplication of two vectors to give a scalar. The scalar product of A and B is defined by $A.B = AB\cos\theta$, where A and B are the magnitudes of A and B and θ is the angle between the vectors. An example is a force F displaced s. Here the scalar product is energy transferred (or work done):

$$W = F.s$$

$$W = Fs\cos\theta$$

where θ is the angle between the line of action of the force and the displacement. A scalar product is sometimes called a *dot product*. *See also* vector product.

scanning electron microscope *See* electron microscope.

scattering The 'spreading out' of a beam of radiation as it passes through matter, reducing the energy moving in the original direction. Depending on the circumstances, scattering can follow any combination of three processes as the radiation interacts with matter particles — reflection (elastic scattering), absorption followed by re-radiation (inelastic or Compton scattering), and diffraction. Thus sunlight is scattered (or diffused) as it passes through cloud and dust in the atmosphere. However, even perfectly clear air scatters sunlight, making the sky colour blue — high frequencies are scattered more than low frequencies.

schlieren technique A method of studying turbulent flow by high-speed photography. If turbulent flow occurs in a fluid, there are localized short-lived regions of low density. These can be recorded by photography because they have a different refractive constant, and appear as streaks on the photograph.

schorl *See* tourmaline.

Schottky defect *See* defect.

scintillation counter A detector of radiation in which radiation produces scintillations (flashes of light) in a phosphor. These are used to produce an electric pulse by means of a photomultiplier. As well as crystalline scintillators various liquids can be used.

sclerotic The hard outer coating of the eyeball. *See* eye.

scotopic vision Vision at low levels of light intensity (night-time vision). Under such conditions the rods in the retina are the main receptors and no colours can be distinguished. *Compare* photopic vision.

screen grid *See* tetrode.

screw A type of machine, related to the inclined plane, and, in practice, to the class two lever. The efficiency of screw systems is very low because of friction. Even so, the force ratio (F_2/F_1) can be very high.
The distance ratio is given by $2\pi r/p$, where r is the radius and p the pitch of the screw (the angle between the thread and a plane at right angles to the barrel of the screw).

search coil A small coil, in which induced currents are used to measure magnetic field strengths. *See* fluxmeter.

Searle's bar An apparatus for measuring the thermal conductivity of a good conductor. The specimen is a long narrow lagged bar heated at one end by a temperature bath or electric coil. It is cooled at the other end by a coil of copper tubing through which a steady stream of water is passed. The temperatures of the input and output water are measured (θ_3 and θ_4 respectively). In this way a temperature gradient is produced along the bar, which is found by measuring the temperatures (θ_2 and θ_3) at two intermediate points a distance l apart. These temperatures are measured by thermocouples or by thermometers placed in holes in the bar. Under steady-state conditions, thermal conductivity is calculated from the equation:
$$kA(\theta_2 - \theta_1)/l = mc(\theta_4 - \theta_3)$$

m being the mass of water passed in unit time, c the specific thermal capacity of the water, and A the bar's cross-sectional area.

second 1. Symbol: s The SI base unit of time. It is defined as the duration of 9 192 631 770 cycles of a particular wavelength of radiation corresponding to a transition between two hyperfine levels in the ground state of the caesium-133 atom.
2. Symbol: " A unit of angle: 1/3600 of a degree.

secondary cell *See* accumulator.

secondary colours The colours formed by mixing pairs of primary colours (hues). The effects can be shown by overlapping the beams from three projectors, each projecting a primary colour.

secondary emission The emission of electrons from a surface as a result of the impact of high-energy electrons or ions. The incident particles must have sufficient energy to eject the 'free' electrons in the solid; i.e. to overcome the work function.

secondary wavelets *See* Huygens' construction.

secondary winding *See* transformer (voltage).

Seebeck effect The production of an e.m.f. in a circuit containing two conductors, when the junctions between the two conductors are at different temperatures. The Seebeck effect is the phenomenon producing the e.m.f. in thermocouples. *Compare* Peltier effect.

selenium cell A photoconducting cell consisting of a piece of light-sensitive selenium semiconductor, which decreases its electrical resistance when light falls on it. It is powered by a battery. *See also* photocell.

self-inductance *See* self-induction.

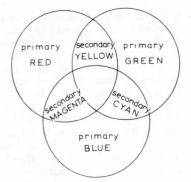

yellow = red + green = white − blue
cyan = green + blue = white − red
magenta = blue + red = white − green

self-induction The production of an e.m.f. in a conductor (e.g. a coil) caused by changes in the current in the conductor itself. The electromotive force is sometimes known as the *back e.m.f.* as its direction opposes the current change in accordance with Lenz's law. The relationship between the e.m.f. (E) and the rate of change of current (dI/dt) is given by the equation:

$$E = L(dI/dt)$$

where L is a constant for the coil known as its *self-inductance*. The SI unit of inductance is the henry (H). *Compare* mutual induction.

semiconductor A material, such as silicon or germanium, that has a resistivity midway (on a logarithmic scale) between that of conductors and that of insulators. In a pure semiconductor material, called an *intrinsic semiconductor*, the concentrations of negative charge carriers (electrons) and positive charge carriers (holes) are the same. In practice, absolutely pure semiconductor does not exist, but the intrinsic conductivity is used as a reference level for an impure semiconductor. The conductivity increases considerably when certain impurities are added, depending strongly on the type and concentration of the impurity. Such a material is called an *extrinsic semiconductor*. The process of adding impurity to control the conductivity is called *doping*.

Addition of, for example, phosphorus increases the number of electrons available for conduction, and the material is described as n-type, i.e. the charge carriers are negative. The impurity, or *dopant*, is called a *donor* impurity in this case.

Addition of, for example, boron has the effect of removing electrons; the dopant in this case being called an *acceptor* impurity because its atoms accept electrons for the completion of atomic bonds, leaving behind positive holes. The material is described as p-type, i.e. the charge carriers are positive; the holes are regarded as mobile.

Holes in p-type and electrons in n-type semiconductors are described as *majority carriers*; there are also electrons in p-type and holes in n-type semiconductors, but they are in the minority and are called *minority carriers*.

Semiconductors have very many uses. They distinguish solid-state devices from vacuum thermionic valves, now mostly superseded by transistors. *See also* diode, integrated circuit, transistor.

semiconductor counter A device for detecting ionizing radiation, consisting of a semiconductor crystal with a potential difference across it. Particles striking the crystal produce electron–ion pairs, which increase the conductivity of the crystal for a brief period. Pulses of current are thus produced, and can be counted.

semiconductor diode *See* diode.

series Elements in a circuit are *in series* if connected so that each carries the same current in turn.

For resistors in series, the resulting resistance is given by:

$$R = R_1 + R_2 + R_3 + \dots$$

For cells in series, the resulting e.m.f. is given by:

$$E = E_1 + E_2 + E_3 + \dots$$

For capacitors in series, the resulting capacitance is given by

$$1/C = 1/C_1 + 1/C_2 + 1/C_3 + \dots$$

Compare parallel.

series-wound motor An electric motor in which the field winding is connected in series with the armature winding. Series-wound motors are used for applications in which a high starting torque is required (e.g. in cranes). Initially the current is very high through both armature and field coil, and the torque is large. Series-wound motors, however, are less suitable than shunt-wound motors for maintaining steady speeds. *Compare* shunt-wound motor.

sextant A navigational instrument used for the accurate measurement of the angle of a heavenly body above the horizon. Two mirrors are used: one sights the object; the other, part silvered, reflects an image of that object as well as showing the horizon.

shadow The result of the interception of light by an opaque obstacle. As light rays travel in straight lines, light from the source cannot be detected behind the obstacle. A point source of light will produce a very sharp dark shadow behind an obstacle. This is called an *umbra* (total shadow). A larger source will also produce an umbra behind an obstacle. However this will be surrounded by a *penumbra* (partial shadow). Each point in the penumbra will receive some light, so will not be in total shadow. Exactly the same ideas apply to shadows formed in other radiations.

shear modulus *See* elastic modulus.

shear strain *See* strain.

SHF *See* superhigh frequency.

s.h.m. *See* simple harmonic motion.

short circuit The connection of two points in an electric circuit across a very low resistance, so that most of the current bypasses part of the circuit. Short circuits may occur accidentally because of wiring faults.

short sight *See* myopia.

shunt An element connected in parallel with another component and used to take part of the current in the circuit. Thus, a low resistance in parallel with an ammeter, which can be used to

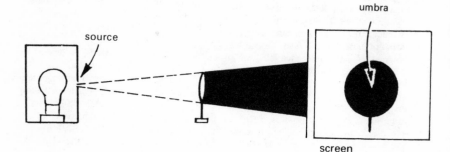

Shadow formation with point source

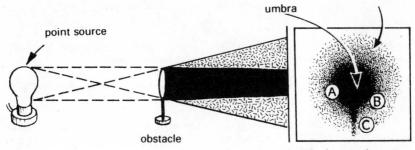

Shadow from an extended source

change the range of the meter, is called a shunt.

shunt-wound motor An electric motor in which the field winding is connected in parallel with the armature winding. Shunt-wound motors are used for maintaining a fairly steady speed (e.g. in machine tools). If the load increases the back e.m.f. falls and the current increases, tending to maintain the speed. *Compare* series-wound motor.

SI *See* SI units.

sideband A range of frequencies contained in a modulated carrier wave, either above or below the unmodulated frequency (hence upper and lower sidebands). The existence of sidebands is a consequence of the modulation process. For instance, in an amplitude modulated wave, if the carrier frequency is f_c and the modulating signal frequency is f_s, then the modulated wave has three components of frequency $f_c - f_s$, f_c, and $f_c + f_s$. *See also* carrier wave, modulation.

sidereal Based on the stars, used especially of measurements.

sidereal time *See* time, day, year.

siemens (mho) Symbol: S The SI unit of electrical conductance, equal to a conductance of one ohm^{-1}.

sigma particle *See* elementary particles.

sign convention A set of rules for determining the signs of distances related by various formulae in optics. In the most common *real is positive, virtual is negative* system, distances to real objects, images, and focal points are positive; distances to virtual points are negative.

The other main sign convention, the *New Cartesian convention*, is now less common than it was. Intended to relate directly to the sign convention in the mathematical treatment of Cartesian coordinates, it treats distances right of the pole as positive; those points to the left of it are negative. Despite the advantage of conformity with mathematical practice, this approach leads to the serious problem that formulae take different forms in different situations.

simple harmonic motion (s.h.m.) Any motion that can be drawn as a sine wave. Examples are the simple oscillation (vibration) of a pendulum or a sound source and the variation involved in a simple wave motion. Simple harmonic motion is observed when the system, moved away from the central position, experiences a restoring force proportional to the displacement from this position.

The equation of motion for such a system can be written in the form:
$$m\mathrm{d}^2x/\mathrm{d}t^2 = -\lambda x$$

Typical forms of simple harmonic motion

System	Period	Symbols	Notes
mass on spring	$2\pi\sqrt{(m/k)}$	k = spring constant	Hooke's law region
uniform 'bobbing' float	$2\pi\sqrt{(h/g)}$	h = immersion at rest	
liquid in U-tube	$2\pi\sqrt{(l/2g)}$	l = column length	
simple pendulum	$2\pi\sqrt{(l/g)}$	l = pendulum length	approximate s.h.m.

λ being a constant. During the motion there is an exchange of kinetic and potential energy, the sum of the two being constant (in the absence of damping). The period (T) is given by:

$$T = 1/f$$

or

$$T = 2\pi/\omega$$

where f is frequency and ω pulsatance. Other relationships are:

$$x = x_0\sin\omega t$$
$$dx/dt = \pm\omega\sqrt{(x_0^2 - x^2)}$$
$$d^2x/dt^2 = -\omega^2 x$$

Here x_0 is the maximum displacement; i.e. the amplitude of the vibration. In the case of angular motion, as for a pendulum, θ is used rather than x.

simple pendulum See pendulum.

sine wave The waveform resulting from plotting the sine of an angle against the angle. Any motion that can be plotted so as to give a sine wave is a simple harmonic motion.

sinusoidal Describing a quantity that has a waveform that is a sine wave.

SI units (Système International d'Unités) The internationally adopted system of units used for scientific purposes. It has seven base units (the metre, kilogram, second, kelvin, ampere, mole, and candela) and two supplementary units (the radian and steradian). Derived units are formed by multiplication and/or division of base units; a number have special names. Standard prefixes are used for multiples and submultiples of SI units. The SI system is a coherent rationalized system of units.

sky wave A radio wave that is transmitted between points by reflection from the ionosphere (a region of electrons and charged particles high in the Earth's atmosphere). Compare ground wave.

slide projector See projectors.

sliding friction See friction.

Snell's law See refraction, laws of.

soft iron A form of iron (containing little carbon) that can easily be magnetized, but does not retain its magnetization when the external field is removed. Many such substances are now known. They are of great value as the core material of electromagnetic machines, such as transformers, generators, and electromagnets.

soft (low) vacuum See vacuum.

solar constant See irradiance.

solar eclipse See eclipse.

solar time See time, day, year.

solenoid A tight coil of wire with a diameter that is small compared with the length. When carrying a current, the solenoid's magnetic field is similar to that of a bar magnet. It can be used as an electromagnet if a core of soft iron is used.

The flux density is given by $\mu_0 NI/l$, where N is the number of turns, I the current, and l the length.

solid See states of matter.

Base and Supplementary SI Units

physical quantity	name of SI unit	symbol for unit
length	metre	m
mass	kilogram(me)	kg
time	second	s
electric current	ampere	A
thermodynamic temperature	kelvin	K
luminous intensity	candela	cd
amount of substance	mole	mol
*plane angle	radian	rad
*solid angle	steradian	sr
*supplementary units		

Derived SI Units with Special Names

physical quantity	name of SI unit	symbol for SI unit
frequency	hertz	Hz
energy	joule	J
force	newton	N
power	watt	W
pressure	pascal	Pa
electric charge	coulomb	C
electric potential difference	volt	V
electric resistance	ohm	Ω
electric conductance	siemens	S
electric capacitance	farad	F
magnetic flux	weber	Wb
inductance	henry	H
magnetic flux density	tesla	T
luminous flux	lumen	lm
illuminance (illumination)	lux	lx
absorbed dose	gray	Gy

Decimal Multiples and Submultiples to be used with SI Units

submultiple	prefix	symbol	multiple	prefix	symbol
10^{-1}	deci-	d	10^1	deca-	da
10^{-2}	centi-	c	10^2	hecto-	h
10^{-3}	milli-	m	10^3	kilo-	k
10^{-6}	micro-	μ	10^6	mega-	M
10^{-9}	nano-	n	10^9	giga-	G
10^{-12}	pico-	p	10^{12}	tera-	T
10^{-15}	femto-	f	10^{15}	peta-	P
10^{-18}	atto-	a	10^{18}	exa-	E

solid angle Symbol: Ω The three-dimensional analogue of angle; it is subtended at a point by a surface (rather than by a line). The unit is the steradian (sr), which is defined analogously to the radian — the solid angle subtending unit area at unit distance. As the area of a sphere is $4\pi r^2$, the solid angle corresponding to the revolution (2π radians) is 4π steradians.

solidification *See* freezing.

solid state Describing electronic devices that use semiconductor circuit elements rather than valves.

sonar (asdic) An acronym for *so*und *na*vigational *r*anging. A technique for locating objects underwater by transmitting a pulse of ultrasonic 'sound' and detecting the reflecting pulse. The time delay between transmission of the pulse and reception of the reflected pulse indicates the depth of the object. An apparatus for determining depths in this way is called an *echo sounder*.

sonometer (monochord) A device for investigating the vibration of a fixed string or wire. Essentially, it is a hollow box with the wire stretched across the top. One end of the wire is fixed; the other passes over a pulley and supports a load, which gives it tension. The length of wire is determined by two inverted V-shaped bridges under the wire. The wire is plucked and the note determined by comparison with tuning forks. In this way the effect of length and tension can be investigated. The frequency (f) produced is given by
$$f = (1/2l)\sqrt{(T/m)}.$$
where l is the length, T the tension, and m the mass per unit length of the string.

sorption Absorption of gases by solids.

sorption pump A type of vacuum pump in which gas is removed from the system by absorption onto a porous solid (e.g. zeolite or charcoal) cooled in liquid nitrogen.

sound Vibrations transmitted by air, or some other material medium, in the form of alternate compressions and rarefactions of the medium. Sound is often called a pressure wave. Sound is a longitudinal waveform, and is characterized by its pitch (frequency), loudness (intensity), and quality. The speed of sound depends on the medium transmitting it. In air at 0°C it is 331.3 metre second^{-1}, in water at 25°C it is 1498 metre second^{-1}, and in glass at 20°C it is 5000 metre second^{-1}. The speed of sound in a medium is given by:
$$V = \sqrt{(E/\rho)}$$
where ρ is the density of the medium and E a modulus of elasticity. For a solid E is the Young modulus; for a liquid it is the bulk modulus. For a gas E is equivalent to the adiabatic bulk modulus; i.e.
$$V = \sqrt{(\gamma p/\rho)}$$
where p is the pressure of the gas and γ the ratio of the molar thermal capacities. This is equivalent to
$$V = \sqrt{(\gamma RT/M)}$$
where R is the molar gas constant, T the thermodynamic temperature, and M the mass of one mole of the gas. V is independent of pressure and depends on the square root of temperature.

source *See* field effect transistor.

south pole *See* poles, magnetic.

space charge A region in a vacuum, gas, or semiconductor in which there is a net electric charge resulting from an excess or deficiency of electrons.

space–time In Newtonian (pre-relativity) physics space and time are separate and absolute quantities; that is they are the same for all observers in any frame of reference. An event seen in one frame is also seen in the same place and at the same time by another observer in a different frame.
After Einstein had proposed his theory of relativity, Minkowski suggested that since space and time could no longer be regarded as separate continua, they should be replaced by a single

continuum of four dimensions, called space-time. In space-time the history of an object's motion in the course of time is represented by a line called the 'world curve'. *See also* frame of reference, relativity.

spark counter A type of detector for counting alpha particles. It consists of a wire (or mesh) held a close distance above an earthed plate. The wire has a high positive potential, less than that required to cause a spark. When an alpha particle passes close to the wire the electric field increases and a spark occurs between the electrodes. The sparks can be counted by a suitable circuit.

spark, electric *See* electric spark.

Special theory *See* relativity.

specific Denoting a physical quantity per unit mass. For example, volume (*V*) per unit mass (*m*) is called specific volume: $V = V/m$. In certain physical quantities the term does not have this meaning: for example, *specific gravity* is more properly called relative density; *specific resistance* is resistivity.

specific gravity *See* relative density.

specific humidity *See* humidity.

specific latent heat *See* latent heat.

specific latent thermal capacity *See* latent heat.

specific resistance *See* resistivity.

specific rotatory power Symbol: α_m The rotation of plane-polarized light in degrees produced by a 10 cm length of solution containing 1 g of a given substance per millilitre of stated solvent. The specific rotatory power is a measure of the optical activity of substances in solution. It is measured at 20°C using the D-line of sodium.

specific thermal capacity (specific heat capacity) Symbol: *c* The energy required to raise the temperature of unit mass of a given substance by unit temperature. Generally, it is given as the energy in joules that raises the temperature of one kilogram by one kelvin; the units are then joules per kilogram per kelvin ($J\,kg^{-1}\,K^{-1}$).

Note that specific thermal capacity is thermal capacity per unit mass, and is a property of the material or substance rather than of a given sample. Water, for example, has a specific thermal capacity of around $4.2\,kJ\,kg^{-1}\,K^{-1}$.

Strictly speaking, specific thermal capacities depend on the way in which pressure and volume change during energy transfer. This is particularly important for gases, for which two principal specific thermal capacities are stated: one at constant pressure (c_p) and one at constant volume (c_v). For an ideal gas, the principal specific heat capacities are related by:
$$c_p - c_v = R$$
where R is the universal gas constant. The ratio c_p/c_v is given the symbol γ (*gamma*). In the case of an ideal gas this depends on the number of degrees of freedom (F) of the molecules according to the relationship $\gamma = 1 + 2/F$.
See also thermal capacity, molar thermal capacity.

spectral emissivity *See* emissivity.

spectral line A particular wavelength of light emitted or absorbed by an atom, ion, or molecule. *See* line spectrum.

spectral series A group of related lines in the absorption or emission spectrum of a substance. The lines in a spectral series occur when the transitions are all between the same energy level and a set of different levels. *See also* Bohr theory.

spectrograph An instrument for producing a photographic record of a spectrum.

spectrometer 1. An instrument for examining the different wavelengths present in electromagnetic radiation. A simple type of spectrometer has a flat rotatable table on which a prism or diffraction grating is placed. Light is directed onto this through a collimator tube, and the transmitted light is observed through a telescope. The telescope can be moved on an angular scale, so that different wavelengths of radiation can be seen.

There are many other types of spectrometer for producing and investigating spectra over the whole range of the electromagnetic spectrum. Often spectrometers are called *spectroscopes*.

2. Any of various other instruments for analysing the energies, masses, etc., of particles. *See* mass spectrometer.

spectrophotometer A form of spectrometer able to measure the intensity of radiation at different wavelengths in a spectrum. Spectrophotometers can be made to work with infrared and ultraviolet radiation as well as with visible radiation.

spectroscope An instrument for examining the different wavelengths present in electromagnetic radiation. *See also* spectrometer.

spectroscopy 1. The production and analysis of spectra. There are many spectroscopic techniques designed for investigating the electromagnetic radiation emitted or absorbed by substances. Spectroscopy, in various forms, is used for analysis of mixtures, for identifying and determining the structures of chemical compounds, and for investigating energy levels in atoms, ions, and molecules. In astronomy it is used for determining the composition of celestial objects and for measuring red shifts.

2. Any of various techniques for analysing the energy spectra of beams of particles or for determining mass spectra.

spectrum 1. A range of electromagnetic radiation emitted or absorbed by a substance under particular circumstances. In an *emission spectrum*, light or other radiation emitted by the body is analysed to determine the particular wavelengths produced. The emission of radiation may be induced by a variety of methods; for example, by high temperature, bombardment by electrons, absorption of higher-frequency radiation, etc. In an *absorption spectrum* a continuous flow of radiation is passed through the sample. The radiation is then analysed to determine which wavelengths are absorbed. *See also* band spectrum, continuous spectrum, line spectrum.

2. In general, any distribution of a property. For instance, a beam of particles may have a spectrum of energies. A beam of ions may have a mass spectrum (the distribution of masses of ions). *See* mass spectrometer.

spectrum, electromagnetic *See* electromagnetic spectrum.

specular reflection *See* reflection.

speed Symbol: c Distance moved in unit time: $c = d/t$. Speed is a scalar quantity; the equivalent vector is velocity, or displacement in unit time.

Usage can be confusing. It is common to meet the word velocity where speed is more correct. For instance c_0 is the speed of light in free space, not its velocity; temperature relates to the root-mean-square speed of particles.

speed of sound *See* sound.

spherical aberration *See* aberration.

spherometer An instrument for measuring the curvature of a spherical surface.

spin A property of certain elementary particles whereby the particle acts as if it were spinning on an axis; i.e. it has an angular momentum. Such particles also have a magnetic moment. In a magnetic field the spins line up at an angle to the field direction and precess around this direction. Certain definite orientations to the field direction occur such that $m_s h/2\pi$ is the component of angular momentum along this direction. m_s is

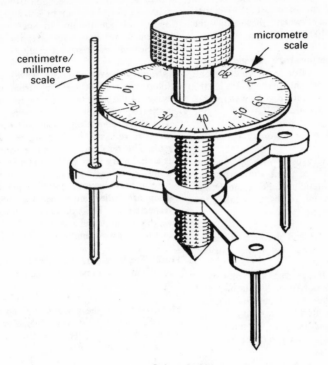

centimetre/
millimetre
scale

micrometre
scale

Spherometer

the spin quantum number. For an electron it has values + 1/2 and -1/2.

S-pole *See* poles, magnetic.

square wave A stream of regular on-off pulses, usually in an electrical signal. The pulses are of equal height and the duration of each pulse is the same as the interval between them.

stability A measure of how hard it is to displace an object or system from equilibrium.
Three cases are met in statics, as shown. They differ in the effect on the centre of mass of a small displacement. An object's stability is improved by: (a) lowering the centre of mass; or (b) increasing the area of support; or by both. *See also* virtual work.

stable equilibrium Equilibrium such that if the system is disturbed a little, there is a tendency for it to return to its original state. *See* stability.

standard cell A voltaic cell whose e.m.f. is used as a standard. *See* Clark cell, Weston cadmium cell.

standard electrode A half cell used for measuring electrode potentials. The hydrogen electrode is the basic standard but, in practice, calomel electrodes are usually used.

standard near-point distance *See* near-point.

standard pressure An internationally agreed value; a barometric height of 760 mmHg at 0°C; 101 325 Pa (approximately 100 kPa).

This is sometimes called the *atmosphere* (used as a unit of pressure). The *bar*, used mainly when discussing the weather, is 100 kPa exactly.

standard temperature An internationally agreed value for which many measurements are quoted. It is the melting temperature of water, 0°C (273.15 K). *See also* STP.

standing wave *See* stationary wave.

stat- A prefix used with a practical electrical unit to name the corresponding electrostatic unit. For example, the electrostatic unit of charge is called the *statcoulomb. Compare* ab-.

states of matter The three states — solid, liquid, and gas — in which matter normally exists.
Solids have fixed shapes and volumes — i.e. they do not flow, like liquids and gases, and they are difficult to compress. In solids the atoms or molecules occupy fixed positions in space. In most cases there is a regular pattern of atoms — the solid is crystalline.
Liquids have fixed volumes (i.e. low compressibility) but flow to take up the shape of the container. The atoms or molecules move about at random, but they are quite close to one another and the motion is hindered.
Gases have no fixed shape or volume. They expand spontaneously to fill the container and are easily compressed. The molecules have almost free random motion.
A plasma is sometimes considered to be a fourth state of matter.

static electricity 1. *See* electrostatics.
2. Electricity characteristic of charges that are at rest, as opposed to current electricity.

static friction *See* friction.

static pressure The pressure on a surface due to a second solid surface or to a fluid that is not flowing.

statics A branch of mechanics dealing with the forces on an object or in a system in equilibrium. In such cases there is no resultant force or torque and therefore no resultant acceleration.

stationary wave (standing wave) The interference effect resulting from two waves of the same type moving with the same frequency through the same region. The effect is most often caused when a wave is reflected back along its own path. The resulting interference pattern is a stationary wave pattern. Here some points always show maximum amplitude; others show minimum amplitude. They are called *antinodes* and *nodes* respectively. The distance between neighbouring node and antinode is a quarter of a wavelength.

stator The fixed part of an electric motor or dynamo, or some similar device. *See also* rotor.

steam Water in the gaseous state above 100°C. Note that the white 'steam' seen when boiling water contains tiny droplets of liquid water.

steam point *See* International Practical Temperature Scale.

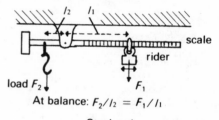

load F_2 F_1

At balance: $F_2/l_2 = F_1/l_1$

Steelyard

steelyard An ancient weighing device, still widely used, as shown or in more complex arrangements. In use, the movable rider is placed so that the device rests horizontally. The weight of the load is then read from the scale directly.

Stefan's constant *See* Stefan's law.

178

Stefan's law The principle that the total energy, W, radiated by the surface of a black body is proportional to the fourth power of the thermodynamic temperature (T). For a surface, $M = \sigma T^4$, where M is the energy radiated per unit area per second. The constant σ is known as the *Stefan constant* and is equal to 5.7 $\times$ 10^{-8} watt metre^{-1} kelvin^{-4} (W m^{-1} K^{-4}). *See also* black body.

step-down transformer A transformer for reducing an alternating voltage.

step-up transformer A transformer for increasing an alternating voltage.

steradian Symbol: sr The SI unit of solid angle. The surface of a sphere, for example, subtends a solid angle of 4π at its centre. The solid angle of a cone is the area intercepted by the cone on the surface of a sphere of unit radius.

stereoscope A device that shows different views of the same scene to each eye in such a way as to give the impression of depth. Stereoscopes were very popular in the last century. Although they can be extremely effective, they are now rare.

stereoscopic vision *See* binocular vision.

stimulated emission If an atom or molecule is in an excited state it may make a transition to a lower-energy state spontaneously, with emission of a photon. Alternatively, this transition may occur by 'impact' of another photon of the required energy — a process known as stimulated emission. It is the basis of the action of the laser. *See* laser.

stokes Symbol: St A former unit of kinematic viscosity used in the c.g.s. system. It is equal to 10^{-4} m^2 s^{-1}.

Stokes' law In fluid flow, the relationship between the frictional force on an object moving through a viscous fluid, and the object's speed, its dimensions, and the fluid properties. For a sphere of radius r falling freely under gravity with

a speed c through a fluid of viscosity η, the frictional force $F = 6\pi r\eta c$. The sphere accelerates until c reaches the steady *terminal speed*, at which F equals the driving force on the sphere. Stokes' law can be used to measure fluid viscosity.

stop An aperture, such as an iris or diaphragm, in an optical system.

storage battery *See* accumulator.

STP (NTP) Standard temperature and pressure. Conditions used internationally when measuring quantities that vary with both pressure and temperature (such as the density of a gas). The values are 101 325 Pa (approximately 100 kPa) and 0°C (273.15 K). *See also* standard pressure, standard temperature.

strain The fractional change in dimension produced by a stress applied to a body. *Tensile strain* applies to the stretching of a body. It is the change in length divided by the original length ($\Delta l / l$). *Bulk strain* occurs when a body is subjected to a pressure. It is the change in volume divided by the original volume. *Shear strain* occurs when an angular deformation occurs, and is equal to the angular displacement produced. *See also* elasticity, elastic modulus.

strain gauge A device for measuring the strain or distortion on a surface. Some strain gauges work by changes in the electrical resistance of a wire attached to the surface. Others use magnetic induction in two coils, one fixed to the surface. Capacitance changes can be used in a similar way.

streamline A line following the direction of the fluid in laminar or *streamline flow*. Where the speed increases, as it does in a narrower section of a pipe, the streamlines are closer together. *See also* laminar flow.

stress A force per unit area applied to a body. *Tensile stress* is a stress tending to stretch a body. *Bulk stress* is an overall force per unit area (pressure) applied. *Shear stress* is a stress tending to produce an angular deformation. It is the tangential force per unit area. *See also* elasticity, elastic modulus.

strobe *See* stroboscope.

stroboscope (strobe) A device that allows something to be viewed as a series of short separate scenes, rather than continuously. It may be a light giving regular short intense flashes, or a rotating disc with slots through which the system is viewed. Stroboscopes are used in physics to investigate periodic motions (rotation or vibration). If the frequency of the illumination is equal to the frequency of vibration, the system is seen at the same point in its cycle at each view, and appears stationary.

strong interaction *See* interaction.

subatomic particles *See* elementary particles.

subcritical Describing an arrangement of fissile material that does not permit a sustained chain reaction because too many neutrons are absorbed without causing fission or otherwise lost. *Compare* supercritical. *See also* multiplication factor.

sublimation A direct change of state from solid to vapour without melting.

subtractive process A process of colour mixing by subtraction. *See* colour.

superconductivity The effective disappearance of electrical resistance in certain substances when they are cooled close to absolute zero. Currents induced in circuits of such materials have persisted for several years with no measurable change. Superconducting circuits cooled by liquid helium are used to produce large magnetic fields for particle accelerators, nuclear fusion experiments, etc. Many other applications are being considered.

supercooling The cooling of a liquid at a given pressure to a temperature below its melting temperature at that pressure without solidifying it. The liquid particles lose energy but do not spontaneously fall into the regular geometrical pattern of the solid. A supercooled liquid is in a metastable state and will usually solidify if a small crystal of the solid is introduced to act as a 'seed' for the formation of crystals. As soon as this happens, the temperature returns to the melting temperature until the substance has completely solidified.

supercritical Describing an arrangement of fissile material that sustains a branching chain reaction; i.e. more neutrons are being produced than are wasted and escape. *Compare* subcritical. *See also* multiplication factor.

superelastic collision A collision for which the restitution coefficient is greater than one. In effect the relative velocity of the colliding objects after the interaction is greater than that before. The apparent energy gain is the result of transfer from potential energy. For example, if a collision between two trolleys causes a compressed spring in one to be released against the other, the collision may be superelastic. *See also* restitution (coefficient of).

superficial expansivity *See* expansivity.

superfluidity A property of liquid helium at very low temperatures. At 2.186 K liquid helium makes a transition to a superfluid state, which has a high thermal conductivity and flows without friction.

superheating The raising of a liquid's temperature at a given pressure above its boiling temperature at that pressure without change of state. This is done by increasing the pressure, as in a pressure cooker, in which the water vapour is not allowed to expand and the pressure inside increases, thus increasing the boiling point.

superhigh frequency (SHF) A radio frequency in the range between 30 GHz and 3 GHz (wavelength 1-10 cm).

superposition, principle of See principle of superposition.

supplementary units The dimensionless units — the radian and the steradian — used along with base units to form derived units. See also SI units.

suppressor grid See pentode.

surface charge density See charge density.

surface tension Symbol: γ or o The attraction between molecules (*cohesion*) at the surface of a liquid, which thus acts a bit like an elastic skin containing the liquid. Surface tension explains why water can drip slowly from a tap and why mercury, gathers into globules on a flat surface. Molecules that are surrounded by others are, on average, pulled equally in all directions. At the surface, however, the net force is inwards, tending to reduce the surface area.

γ is defined as the force acting across the surface per unit of surface length perpendicular to the force. It is measured in units of newton metre^{-1}. Sometimes called *free surface energy*, it may also be defined as the energy involved in increasing the surface area by a square metre. Both are defined at given temperature, as surface tension falls with temperature rise.

susceptibility Symbol: X The ratio, for a given substance, of the magnetization of a sample to the magnetic field strength applied. In SI it equals (μ_r - 1), where μ_r is the relative permeability. The value of X determines whether a substance shows *paramagnetism*, *diamagnetism*, or *ferromagnetism*. A diamagnetic material has a negative susceptibility while paramagnetic and ferromagnetic materials have small and large positive susceptibilities respectively.

synchrocyclotron A device for acceler-

ating particles to high energies; a development of the cyclotron in which the frequency applied to the 'dees' is decreased to compensante for the progressively increased period of revolution caused by the relativistic increase in mass at high energies, typically hundreds of MeV. For example, the 4.7 m machine at Berkeley in the United States can produce particles of energy up to 730 MeV. The synchrocyclotron is also sometimes called a frequency modulated (FM) cyclotron.

synchronous motor An electric motor that runs at a speed proportional to the frequency of the voltage supply. A rotating magnetic field is produced by separate fixed coils energized by an alternating current supply. This field pulls the rotating part round with it. The rotating part of the device consists of a permanent magnet or an electromagnet. When starting, the motor behaves like an induction motor until the synchronous speed is reached. See also electric motor, induction motor.

Système International d'Unités See SI units.

T

tandem generator An arrangement of two linear accelerators end to end. The ends are earthed and there is a common anode at the centre at a potential difference of up to 20 million volts. Negative ions are accelerated from earth potential at one end to the centre where the surplus electrons are stripped off to convert them into positive ions, which then continue to accelerate to the far end. They thus get the acceleration due to the high potential twice over. Protons can be given energies up to 40 MeV.

tangent galvanometer A galvanometer

in which a magnetized needle is pivoted to rotate in a horizontal plane at the centre of a large fixed vertical coil. Current flowing through the coil turns the needle. Tangent galvanometers are now rarely used.

telescope An optical system for collecting radiation from a distant object and producing an enlarged image. *Optical telescopes* use visible radiation. There are two types: *refractors* (or *refracting telescopes*) in which the light is collected by a converging lens; and *reflectors* (or *reflecting telescopes*) in which a converging mirror is used. The light collected by the objective forms an image, which is magnified by an eyepiece. A large objective is desirable in that it collects more light from faint distant objects. A large objective also minimizes diffraction effects.

Reflector systems have a number of advantages over refractors in theory and practice, and the largest *astronomical telescopes* are all reflectors. They do not have the same problems of chromatic aberration; large, comparatively cheap, robust instruments can give good images. Telescopes for non-astronomical use are refractors that produce an upright image (known as *terrestrial telescopes*).

The performance of telescopes is described in terms of:

(1) *magnifying power* — the angular magnification obtained (usually the ratio of focal distance of objective to that of eyepiece).

(2) *aperture* — determining the amount of light admitted, and thus the minimum magnitude of sources that can be seen.

(3) *resolving power* — the ability to separate images of close objects.

Telescopes for use in astronomy have been developed for other forms of electromagnetic radiation.

See also reflector, refractor.

temperature A measure of the 'hotness' of a body. Temperature measures the energies of the particles in a sample. In practice, temperature is measured by using some physical property that depends in a known way on temperature

(e.g. resistance, volume, etc.) and using a defined scale.

Temperature can be regarded as the property that determines whether there can be net heat flow between two bodies. According to the kinetic theory, gas temperature is determined by the mean square speed of the molecules:

$$(Nmc^2)/3 = RT$$

See also temperature scale, thermometer, thermodynamic temperature.

temperature coefficient A coefficient determining how some physical property changes with temperature. For example, the temperature coefficient of resistance is the coefficient in the equation showing how resistance or resistivity varies with temperature. *See* resistance thermometer.

temperature scale A practical scale for use when measuring temperature. A temperature scale is determined by choosing fixed temperatures (*fixed points*), which are reproducible systems assigned an agreed temperature. For example, on the Celsius scale the two fixed points were the temperature of pure melting ice (the *ice temperature*) and the temperature of pure boiling water (the *steam temperature*). The difference between the fixed points is the *fundamental interval* of the scale. The fundamental interval is subdivided into temperature units. In the case of the Celsius scale, the fundamental interval is divided into 100 degrees. The International Temperature Scale has 11 fixed points covering the range 13.81 kelvin to 1337.58 kelvin.

temporary magnetism Substances show temporary magnetism if any magnetism disappears as soon as the inducing field is removed. They are easy to magnetize, and are often called 'soft', as in 'soft iron'.

tensile strain *See* strain.

tensile strength The tensile *stress* at which a material breaks apart or deforms permanently.

tensimeter A device that measures the difference in vapour pressures between two liquids.

tensometer A portable machine for measuring the tensile strength and other mechanical properties of materials.

tera- Symbol: T A prefix denoting 10^{12}. For example, 1 terawatt (TW) = 10^{12} watts (W).

terminal A point in an electric circuit to which a connecting lead can be attached.

terminal speed The steady final speed reached by a body in a fluid when the resultant force on it is zero.

terrestrial magnetism *See* Earth's magnetism.

terrestrial telescope A refracting telescope for non-astronomical use: one producing an erect image. Terrestrial telescopes are either Galilean telescopes or Keplerian telescopes with added inverting lenses. *See also* Galilean telescope, Keplerian telescope, refractor.

tesla Symbol: T The SI unit of magnetic flux density, equal to a flux density of one weber of magnetic flux per square metre. $1\,T = 1\,Wb\,m^{-2}$.

tesla coil A device for producing high voltage oscillations from a low voltage direct current source. It consists of a transformer with a high ratio of secondary to primary turns and a make-and-break mechanism to interrupt the current in the secondary circuit.

tetrode A thermionic valve with one grid more than the triode, called the *screen grid*. It is placed between the control grid and the anode, and is normally kept at a fixed positive potential. The purpose of the screen grid is to minimize the capacitance between the control grid and the anode and thus improve the high-frequency performance of the triode as an amplifier or oscillator.

A disadvantage of the tetrode is that, at high anode voltages, secondary electrons emitted from the anode are captured by the screen grid, causing a drop in the anode current. This is overcome in the pentode. *See also* pentode, thermionic valve, triode.

theodolite An optical instrument (using prisms and lenses) for measuring horizontal and vertical angles with great accuracy. It is an essential tool in surveying.

theorem of parallel axes If I_0 is the moment of inertia of an object about an axis, the moment of inertia I about a parallel axis is given by:
$$I = I_0 + md^2$$
m is the mass of the object and d is the separation of the axes.

therm A unit of heat energy equal to 10^5 British thermal units (1.055 056 joules).

thermal capacity (heat capacity) Symbol: C The energy required to raise the temperature of a body by one unit. It is usually in joules per kelvin ($J\,k^{-1}$). The thermal capacity depends on the pressure. For liquids and solids the effect is small, but for gas samples it is necessary to consider two thermal capacities: one at constant pressure (C_p) and one at constant volume (C_v).
See also molar thermal capacity, specific thermal capacity.

thermal conductivity *See* conductivity (thermal).

thermal diffusion A method of separating gas molecules of different masses by maintaining one end of the gas at a lower temperature than the other (i.e. producing a temperature gradient along a column of gas) — the more massive molecules tend to stay at the low-temperature end. It can be used for the separation of isotopes.

thermal equilibrium A state in which there is no net flow of heat. If two bodies are in thermal equilibrium, then they have the same temperature. *See also* equilibrium.

thermal expansion *See* expansion, thermal.

thermal neutron *See* moderator.

thermal radiation *See* infrared, radiation.

thermal reactor *See* nuclear reactor.

thermionic diode *See* diode.

thermionic emission The emission of electrons from the surface of a heated metal or metal oxide. The energy of the emitted electrons is relatively low, and their rate of emission increases with the temperature of the surface. The thermal energy of the electrons is sufficient to eject them against the counter-attraction of the surface atoms. Heated cathodes are used as sources of electrons in thermionic valves and electron guns.

thermionic valve A device that uses the thermionic effect and exhibits the action of a valve, i.e. current flow in one direction only. In the simplest type, a diode valve, the electron-emitting cathode surface is contained in an evacuated glass envelope together with the second electrode, the anode. Electrons emitted from the cathode by thermionic emission, flow towards the anode if it is made positive with respect to the cathode. This constitutes a current. Reversal of the potential causes the current to fall to zero. There is no current, or very little, in the reverse direction because emission is in one direction only.
There are many other types of vacuum thermionic valve incorporating additional electrodes, called *grids*. Some of these types may be filled with gas at low pressure, in which case the valve is called a thyratron. For many of its original uses the thermionic valve has been replaced by solid-state transistors, but it is still used in special cases such

as high-power amplification (for radio transmission), cathode-ray tubes, and the klystron and magnetron oscillators. *See also* grid, pentode, tetrode, thermionic emission, triode.

thermistor An electrical component with a high temperature coefficient of resistance (i.e. the electrical resistance changes rapidly with temperature). Thermistors are usually semiconductor devices; in one type the resistance decreases with temperature rise. They can be used as sensitive resistance thermometers, which operate at very low temperatures. Another use is in electronic circuits in which it is desirable that the current should increase to a maximum value slowly.

thermocouple A pair of wires of different conductors, welded or soldered at one end, used to measure temperature. A small e.m.f. is produced at the junction between the metals, and this changes with temperature. The thermocouple (or *thermojunction*) can thus be used as a *thermoelectric thermometer*.
Thermocouples are convenient and have low thermal capacity. In inaccurate work a single junction can be used with a galvanometer in the circuit to measure the current produced by the thermoelectric e.m.f. In more accurate work two identical junctions are connected in series. One junction is maintained at a constant low temperature and the other used for temperature measurement. A potentiometer is used to measure the e.m.f.
See also thermometer, Seebeck effect, thermopile, International Practical Temperature Scale.

thermodynamic equilibrium *See* equilibrium.

thermodynamics The study of heat and other forms of energy and the various related changes in physical quantities such as temperature, pressure, density, etc.
The *first law of thermodynamics* states that the total energy in a closed system

is conserved (constant). This means that energy cannot be created or destroyed, but in all processes is simply converted from one form to another, or transferred from one system to another.

A mathematical statement of the first law is:

$$\delta Q = \delta U + \delta W$$

Here, δQ is the heat transferred to the system, δU the change in internal energy (resulting in a rise or fall of temperature), and δW is the external work done by the system. A gas for example may expand at constant pressure, in which case $\delta W = p\delta V$.

The *second law of thermodynamics* can be stated in a number of ways, all of which are equivalent. One is that heat cannot pass from a cooler to a hotter body without some other process occurring. Another is the statement that heat cannot be totally converted into mechanical work — i.e. a heat engine cannot be 100% efficient.

The *third law of thermodynamics* states that the entropy of a substance tends to zero as its thermodynamic temperature approaches zero.

Often a *zeroth law of thermodynamics* is given: that if two bodies are each in thermal equilibrium with a third body, then they are in thermal equilibrium with each other. This is considered to be more fundamental than the other laws because they assume it. *See also* Carnot cycle, entropy.

thermodynamic temperature Symbol: T A temperature measured in kelvins. The size of the kelvin is the same as that of the degree Celsius, and temperatures on the Celsius scale can be converted to thermodynamic temperatures by adding 273.15.

The thermodynamic temperature scale is independent of the working substance. It is based on the principles of a reversible heat engine. A temperature ratio on the thermodynamic scale is defined as the ratio of heat absorbed to heat rejected in a reversible Carnot cycle: $T_2/T_1 = -q_2/q_1$. The zero point is the temperature of the cold reservoir for a perfectly efficient heat engine. The ther-

modynamic scale of temperature (defined by Kelvin) is equivalent to the absolute temperature scale, based on the behaviour of an ideal gas. In practice measurements of thermodynamic temperature use the International Practical Temperature Scale.

thermoelectric effects Effects relating thermal energy and electricity. *See* Peltier effect, Seebeck effect, Thomson effect.

thermoelectric thermometer *See* thermocouple.

thermojunction *See* thermocouple.

thermoluminescent dating A method for dating the firing of samples of pottery. As a result of absorbed alpha radiation, electrons become trapped at energy levels higher than normal. If the temperature is raised, they revert to the ground state with emission of photons (light). The amount of light emitted can be measured. This depends on the number of trapped electrons, which in turn depends on the lapse of time since the pottery was fired. It also depends on the amount of irradiation by alpha particles during that period and the type of material.

thermometer Any instrument or apparatus used for measuring temperature. Thermometers depend for their action on a *thermometric property* — i.e. a physical property that changes in a known way with temperature. For instance, a mercury thermometer depends on the expansion of a thin column of mercury in a capillary tube. The material or substance used for its thermometric property is the *working substance* of the thermometer. *See also* pyrometer.

thermometer, clinical *See* clinical thermometer.

thermometer, maximum and minimum *See* maximum-and-minimum thermometer.

185

thermometric property *See* thermometer.

thermometry The branch of physics concerned with methods of temperature measurement.

thermonuclear reaction *See* fusion.

thermopile A radiant energy detector or meter. It consists of a series of thermocouples, usually made of antimony and bismuth. Radiation strikes the blackened hot junctions, while the cold junctions are kept shielded. Measurement of the thermoelectric e.m.f. produced, enables the temperature of the hot junction to be calculated and hence the total radiation received from the source.

thermostat An instrument used for maintaining temperature within certain limits. It uses a device that cuts off the power when the required temperature is exceeded, and reconnects the supply when the temperature falls. A common method is to exploit the expansion and contraction of a substance with temperature. *See also* bimetallic strip.

Thomson effect The production of an electric potential gradient along a conductor as a result of a temperature gradient along it. Points at different temperatures in the conductor have different electrical potentials.

thorium series *See* radioactive series.

thoron *See* emanation.

timbre *See* quality of sound.

time Measurement of duration between events. The SI unit, the second, is based on vibration of radiation absorbed by caesium atoms. Time is based on astronomical measurements. *Solar time* is measured with respect to the Sun. *Sidereal time* is based on measurement with respect to the stars. *Lunar time* is based on the lunar month (the time between successive New Moons).

tint The colour sensation of mixing a hue (a spectral colour) with white. This reduces the saturation of the hue.

ton An Imperial unit of mass equal to 2240 lbs (1016.047 kg). Sometimes it is called the *long ton* to distinguish it from the U.S. *short ton* (2000 lbs).

tone *See* note.

tonne (metric ton) Symbol: t A unit of mass equal to 10^3 kilograms.

torque Symbol: *T* A turning force (or moment). The torque of a force *F* about an axis (or point) is *Fs*, where *s* is the distance from the axis to the line of action of the force. The unit is the newton metre. Note that the unit of *work*, also the newton metre, is called the joule. Torque is *not* however measured in joules. The two physical quantities are not in fact the same. Work (a scalar) is the scalar product of force and displacement. Torque is the vector product and is a vector at 90° to the plane of the force and displacement. *See also* couple, moment.

torr A unit of pressure equal to a pressure of 101 325/760 pascals (133.322 Pa). It is equal to the mmHg.

torsion Angular strain produced by applying torque (twisting force) to a body, as occurs when a rod or wire is fixed at one end and rotated at the other.

torsional wave A wave motion in which the vibrations in the medium are rotatory simple harmonic motions around the direction of energy transfer.

torsion balance A device for measuring very small forces by the torsion they cause in a wire or fibre. The angle of twist is usually measured by the displacement of a beam of light reflected from an attached mirror (i.e. by an optical lever).

total eclipse *See* eclipse.

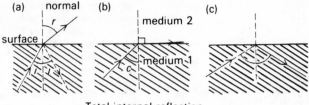

(a) normal (b) (c)

surface

medium 2

medium 1

Total internal reflection

total internal reflection The total reflection of radiation that can occur in a medium at the boundary with another medium of lower refractive constant. In such cases, rays incident at small angles will be refracted 'away from the normal'; in other words, the angle of refraction (r) is greater than the angle of incidence (i). Because r is greater than i, it is possible to increase i to a value at which r is 90° (or just under). That value of i is called the *critical angle c*. The relation for the refractive constant for radiation entering this medium is:

$$_1n_2 = 1/\sin c$$

If i is increased still further, refraction cannot occur. The radiation must then all be reflected. Normally, when radiation meets the boundary between two media some will be reflected and some will be refracted. Total internal reflection is the only exception — here *all* the energy is reflected. Because of this, optical instruments often include totally internally reflecting prisms rather than mirrors. The critical angles of optical glasses at visible wavelengths are typically 40° or less.

total radiation pyrometer An instrument used for measuring high temperatures. Radiation from the source is focused by a concave mirror onto a blackened foil to which is attached a thermocouple. The thermoelectric e.m.f. produced can be measured and hence the temperature of the foil. The temperature of the radiating source may then be calculated. *See also* optical pyrometer.

tourmaline (schorl) A complex bluish or brownish mineral important for its polarizing effect on light. It is a dichroic material.

transformer, voltage A device for changing the voltage of an alternating supply. Transformers have no moving parts and operate by the current in one coil, the *primary* winding, electromagnetically inducing a current in another, the *secondary* winding, which forms part of a separate electrical circuit. The ratio of the voltages in the primary and secondary circuits, V_1/V_2, is approximately equal to the ratio of the number of turns in the primary and secondary coils. In an ideal transformer there is no power loss and the primary to secondary current ratio, I_1/I_2, is the inverse of the voltage ratio; i.e. $V_1I_1 = V_2I_2$. In practice, some power is lost because of eddy currents, hysteris, heating in the coils, and incomplete linking of the magnetic flux between the coils. Usually the primary and secondary windings are wound around a laminated iron core designed to give maximum flux linkage. With careful design, transformers can be more than 98% efficient.

Step-up transformers are used to increase voltage and reduce the current of the output from power stations so that losses in transmission lines are minimized. The voltage is reduced in stages by *step-down transformers* nearer the user.

transient A sudden brief disturbance of a system.

transistor A device incorporating two junctions between n-type and p-type semiconductors, and having the property

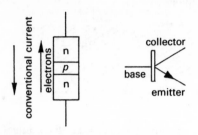

n-p-n transistor

that a current flowing across one junction modulates the current flowing across the other junction. Electrical contact is made by three electrodes attached one to each piece of semiconductor, called the emitter, base, and collector. There are two types of transistor, called p-n-p and n-p-n.

Both types may be regarded as two diodes connected back to back, so that when a voltage is applied across the device one diode is forward biased and the other reverse biased. Current flows across the forward biased junction, i.e. majority carriers (electrons in n-type and holes in p-type semiconductors) move from the emitter towards the collector. The base region is thin enough not to prevent this.

In the n-p-n transistor, electrons from the emitter cross into the base, and some holes from the base (p-type) cross into the emitter. Those electrons entering the base move slowly towards the collector-base junction, which is reverse biased. The base region is relatively thin, and once electrons enter the collector region they are swept through it by the applied voltage. In the opposite direction holes cross the base to the emitter junction, causing the hole concentration in the base to fall. This effect would reduce the forward voltage sufficiently to stop the forward current unless the base region was made to increase its hole concentration by losing electrons through the external connec-

tion to the base. Hence the base-emitter junction must always be forward biased. In such a case current amplification is possible; a small base current controls a larger collector current. The current gain is defined by the ratio of collector current to base current, given by the symbol β (beta). This use of the transistor is called the *common emitter* connection, in which an input signal is applied to the base. This is the most common way in which a transistor is used as an amplifier. *See also* semiconductor.

transition temperature A temperature at which some definite physical change occurs in a substance. Examples of such transitions are change of state, change of crystal structure, and change of magnetic behaviour.

translation *See* translatory motion.

translatory motion (translation) Motion involving change of position; it compares with rotatory motion (rotation) and vibratory motion (vibration). Each is associated with kinetic energy. Translatory motion is usually described in terms of (linear) speed or velocity and acceleration.

translucent Able to pass radiation, but with much deviation and/or absorption. *Compare* transparent.

transmission electron microscope *See* electron microscope.

transmitter A device used in a telecommunication system to generate and propagate an electrical signal.

transparent Able to pass radiation without significant deviation or absorption. Note that a substance transparent to one radiation may be opaque to another. The divide between transparency and translucency is not well defined. Thus some people call a filter transparent (as it does not distort radiation): others would call it translucent (as it absorbs some of the radiation). *See also* opaque, translucent.

transuranic elements Elements that have a proton number greater than 92 (i.e. greater than that of uranium). The transuranic elements are all unstable (radioactive) and are produced by nuclear reactions induced by bombarding heavy elements with high-energy particles.

transverse wave A wave motion in which the motion or change is perpendicular to the direction of energy transfer. Electromagnetic waves and water waves are examples of transverse waves. *Compare* longitudinal waves.

travelling wave *See* wave.

triangle of forces *See* triangle of vectors.

triangle of vectors A triangle describing three coplanar vectors acting at a point with zero resultant. When drawn to scale — shown correctly in size, direction, and sense, but not in position — they form a closed triangle. Thus three forces acting on an object at equilibrium form a *triangle of forces*. Similarly a *triangle of velocities* can be constructed.

triangle of velocities *See* triangle of vectors.

triatomic Describing a molecule consisting of three atoms. Water (H_2O) and carbon dioxide (CO_2) are examples.

triboelectricity (frictional electricity) Static electricity produced by friction.

tribology The study of friction, lubricants, and lubrication.

triboluminescence *See* luminescence.

triode A thermionic valve with three electrodes; i.e. one grid, called the *control grid*, in addition to the anode and cathode. The grid is placed nearer to the cathode than the anode, and so is able to control the current by the application of relatively small voltages to this grid. In normal operation the grid is biased negatively, so that no current flows on to it; but a small voltage fluctuation, or input, superimposed on the grid bias causes large anode current changes, or output. The ratio of output voltage change to input voltage change is called the voltage gain, or *amplification factor*. *See also* grid, grid bias, thermionic valve.

triple point The only point at which the gas, solid, and liquid phases of a substance can coexist in equilibrium. The triple point of water (273.16 K at 101 325 Pa) is used to define the kelvin.

tritiated *See* tritium.

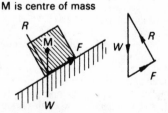

M is centre of mass

(a) Force diagram for an object at rest on a rough slope

(b) Corresponding triangle

Triangle of vectors

tritium Symbol: T A radioactive isotope of hydrogen of mass number 3. The nucleus contains 1 proton and 2 neutrons. Tritium decays with emission of low-energy beta radiation to give ^{3}He. The half-life is 12.3 years. Tritium is found in the atmosphere, possibly as a result of the bombardment of nitrogen by neutrons from cosmic rays:
$$^{14}N + {}^1n \rightarrow {}^{12}C + {}^3H$$
It is useful as a tracer in studies of chemical reactions. Compounds in which ^{3}H atoms replace the usual ^{1}H atoms are said to be *tritiated.*

triton A nucleus of a tritium atom, consisting of two neutrons and one proton.

tuning fork A metal fork designed to vibrate when struck gently to produce a pure sound tone.

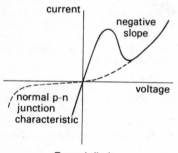

Tunnel diode

tunnel diode A highly doped p-n junction diode that has a large reverse current, and, in the forward direction, a negative slope resistance over part of the voltage–current characteristic.
The 'tunnel effect' explains the shape of the characteristic. Electrons are able to tunnel through the conduction band, which is normally forbidden to them because they do not possess sufficient energy to cross the potential barrier of the junction. *See also* diode, semiconductor.

tunnel effect The passage of a particle through a potential barrier, even though it has not enough energy to pass the barrier on classical grounds. The tunnel effect can be explained by quantum mechanics.

turbulent flow Fluid flow in which the speed at any point varies rapidly in magnitude and direction. Fluid flow becomes turbulent when its speed increases beyond a *critical speed.* This corresponds to a critical value of the Reynolds number that depends on the geometry of the system. *Compare* laminar flow. *See also* Reynolds number.

turns ratio The ratio of the number of turns in the primary coil of a voltage transformer to the number in the secondary coil.

U

UHF *See* ultrahigh frequency.

ultracentrifuge A high-speed centrifuge used for separating out very small particles. The sedimentation rate depends on the particle size, and the ultracentrifuge can be used to measure the mass of colloidal particles and large molecules (e.g. proteins). *See also* centrifuge.

ultrahigh frequency (UHF) A radio frequency in the range between 3 GHz and 0.3 GHz (wavelength 10 cm-1 m).

ultrahigh vacuum *See* vacuum.

ultramicroscope A modified compound microscope able to view the sub-microscopic particles in a colloidal suspension when this is suitably illuminated.

ultrasonic frequency A sound frequency above the range normally audible by humans, i.e. greater than about 20 KHz.

ultrasonics The study of sounds that are

too high-pitched to be detected by the human ear; i.e. have a frequency greater than about 20 kilohertz. Ultrasound is used in a similar way to X-rays, for medical diagnosis and testing for flaws in metal. It is also used to clean surfaces and to break up particles in suspension. Ultrasound is often generated by the piezoelectric effect.
See also sonar.

ultraviolet (UV) A form of electromagnetic radiation, shorter in wavelength than visible light. Ultraviolet wavelengths range between about 1 nm and 400 nm. Ordinary glasses are not transparent to these waves; quartz is a much more effective material for making lenses and prisms for use with ultraviolet.

Like light, ultraviolet radiation is produced by electronic transitions between the outer energy levels of atoms. The distinction between the two types of radiation is in fact physiological rather than physical. However, having a higher frequency, ultraviolet photons carry more energy than those of light.
See also electromagnetic spectrum.

ultraviolet catastrophe In the late nineteenth century it was realized that the short-wavelength region of blackbody radiation could not be explained by the theories of physics of the time (classical physics). The problem — sometimes called the ultraviolet catastrophe — was resolved by the concept of quantization of energy. *See* Planck's radiation law.

umbra Total shadow. *See* shadow.

underdamping *See* damping.

uniaxial crystal A type of birefringent crystal having only one axis along which the ordinary and extraordinary rays travel at the same speed. *See* optic axis.

unified field theory A single theory to account for the electromagnetic, gravitational, strong, and weak interactions by one set of equations. So far, attempts to find such a theory have been unsuccessful, although there has been some progress in unifying the weak and electromagnetic interactions.

uniform acceleration Constant acceleration.

uniform motion A vague phrase, usually taken to mean motion at constant velocity, often in a straight line.

uniform speed Constant speed.

uniform velocity Constant velocity, describing motion in a straight line with zero acceleration.

unipolar transistor *See* field effect transistor.

unit A reference value of a quantity used to express other values of the same quantity. *See also* SI units.

universal constants *See* fundamental constants.

unsaturated Describing a colour that is a tint; i.e. a hue mixed with white. *See* saturation.

unstable equilibrium Equilibrium such that if the system is disturbed a little, there is a tendency for it to move further from its original position rather than to return. *See* stability.

upthrust An upward force on an object in a fluid. In a fluid in a gravitational field the *pressure* increases with depth. The pressures at different points on the object will therefore differ; the resultant is vertically upward. *See* Archimedes' principle, buoyancy, flotation (law of).

uranium-lead dating A method of radioactive dating used for estimating the age of certain rocks. It is based on the decay of ^{238}U to ^{206}Pb (half-life 4.5 $\times$ 10^9 years) or ^{235}U to ^{207}Pb (half-life 0.7 $\times$ 10^9 years). The technique is useful for time periods from 10^7 years ago back to the age of the Earth (about 4.6 $\times$ 10^9 years). *See also* dating.

uranium series *See* radioactive series.

UV *See* ultraviolet.

V

vacancy *See* defect.

vacuum A space containing gas below atmospheric pressure. A perfect vacuum contains no matter at all, but for practical purposes *soft* (*low*) *vacuum* is usually defined as down to about 10^{-2} pascal, and *hard* (*high*) *vacuum* as below this. *Ultrahigh vacuum* is lower than 10^{-7} pascal.

valence band *See* energy bands.

valve voltmeter *See* electrometer.

Van de Graaff accelerator A particle accelerator in which the high voltage from a Van de Graaff generator is used to accelerate charged particles.

Van de Graaff generator An electrostatic generator capable of producing high p.d.'s (up to millions of volts). It consists of a large smooth metal sphere on top of a hollow insulating cylinder. An endless insulating belt runs between pulleys at each end of the cylinder. Charge is sprayed from metal points connected to a high voltage source on to the bottom of the belt. It is then carried up to the top of the belt where it is collected by other metal points and accumulated on the outside of the sphere.

van der Waals equation An equation of state for real gases. For one mole of gas the equation is
$$(p + a/V_m^2)(V_m - b) = RT,$$
where p is the pressure, V_m the molar volume, and T the thermodynamic temperature. a and b are constants for a given substance and R is the gas constant. The equation gives a better description of the behaviour of real gases than the perfect gas equation ($pV_m = RT$).

The equation contains two corrections: b is a correction for the non-negligible size of the molecules; a/V_m^2 corrects for the fact that there are attractive forces between the molecules, thus slightly reducing the pressure from ideal.

van der Waals forces Attractive forces existing between molecules. These forces are the ones giving the pressure correction in the van der Waals equation. They are much weaker than chemical bonds and act over short range (inversely proportional to the seventh power of distance). They are caused by attraction between dipoles of molecules. For atoms or molecules without permanent molecular dipole moments, the attractive forces result from attractions between nucleus–electron dipoles (called dispersion forces).

vaporization A change of state from liquid or solid to vapour.

vapour A gas at any temperature at which it may be liquefied by pressure alone; i.e. a gas below its critical temperature. The critical temperatures for water, carbon dioxide, and nitrogen are 374°C, 31.1°C, and -147°C respectively. *See also* critical state, Andrews' experiments.

vapour pressure The pressure exerted at a particular temperature by a vapour. When a liquid or solid evaporates, molecules are continuously escaping from the surface at a rate that increases rapidly with temperature; they exert a vapour pressure. Those striking the surface tend to re-enter the liquid or solid, so that eventually a state of dynamic equilibrium is reached in which the number of molecules returning to the surface per second is the same as the number leaving it. The vapour now exerts its maximum pressure for that temperature, the *saturated vapour pres-*

sure (SVP). The value of this depends only on the temperature, and is independent of the volume (unless all the vapour condenses or all the liquid evaporates).

variation, magnetic _See_ magnetic variation.

variometer A variable inductor, usually consisting of two coils connected in series and able to be moved with respect to each other, so that the total self-inductance can be changed.

vector A measure in which direction is important and must usually be specified. For instance, displacement is a vector quantity, whereas distance is a scalar. Weight, velocity, and magnetic field strength are other examples of vectors — they are each quoted as a number with a unit and a direction. The addition of vectors must take account of direction. The resultant may be found by the parallelogram of vectors.

There are various ways of multiplying vectors. A vector multiplied by a scalar gives another vector; for example mass (a scalar) multiplied by velocity (a vector) gives momentum (a vector).

Two vectors can be multiplied together in two different ways. _See_ scalar product, vector product.

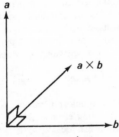

vector product

vector product A multiplication of two vectors to give a vector. The vector product of A and B is written $A \times B$. It is a vector of magnitude $AB\sin\theta$, where A and B are the magnitudes of A and B

and θ is the angle between A and B. The direction of the vector product is at right angles to A and B. It points in the direction in which a right-handed screw would move turning from A towards B. Note, $A \times B = -B \times A$. An example of a vector product is the force F on a moving charge Q in a field B with velocity v (as in the motor effect). Here
$$F = QB \times v$$
The vector product is sometimes called the _cross product_. _See also_ scalar product.

vectors, parallelogram (law) of _See_ parallelogram of vectors.

vectors, triangle (law) of _See_ triangle of vectors.

velocity Symbol: v Displacement per unit time. The unit is the metre per second (m s^{-1}). Velocity is a vector, speed being the scalar form. If velocity is constant, it is given by the slope of a position/time graph, and by the displacement divided by the time taken. If it is not constant, the mean value is obtained. If x is the displacement, the instantaneous velocity is given by $v = dx/dt$. _See also_ equations of motion.

velocity modulation A type of modulation in which the velocity of a beam of electrons is made to fluctuate at the frequency of a superimposed electric field, usually a radiofrequency. As faster electrons overtake slower electrons, bunching occurs in the beam. High-frequency oscillators, such as those used in radar, make use of velocity-modulated beams of electrons. _See also_ modulation, radar.

velocity ratio _See_ distance ratio.

very high frequency (VHF) A radio frequency in the range between 300 MHz and 30 MHz (wavelength 1 m–10 m).

very low frequency (VLF) A radio frequency in the range between 30 kHz and 3 kHz (wavelength 10 km–100 km).

VHF *See* very high frequency.

vibrating-reed electrometer *See* electrometer.

vibration (oscillation) Any regularly repeated to-and-fro motion or change. Examples are the swing of a pendulum, the vibration of a sound source, and the change with time of the electric and magnetic fields in an electromagnetic wave.

vibration magnetometer *See* magnetometer.

virtual image *See* image.

virtual object *See* object.

virtual particle An elementary particle postulated to exist in interactions. For example, it is possible to explain the electromagnetic interaction between two charged particles by assuming that they are exchanging virtual photons.

virtual work The work done if a system is displaced infinitesimally from its position. The virtual work is zero if the system is in equilibrium.

viscometer An instrument for measuring fluid viscosity. Some types measure the rate of fluid flow through a narrow tube. Others measure the time a ball takes to fall through a known depth of fluid. Effects such as the damping of mechanical vibrations by the fluid and the friction forces it transmits between two moving parts (both of which are direct results of viscosity) can also be used in viscometers.

viscosity Symbol: η The resistance of a fluid to flow. Syrup, for example, has a higher viscosity than water. It exerts a greater frictional force when it is stirred, the same object will fall through it more slowly than through water, and it flows more slowly.

In steady flow through a long circular cross-section pipe, the frictional force, F, is proportional to the cross-sectional area, A, times the velocity gradient, dv/dr, (the velocity difference between two points divided by their distance apart). The viscosity, also called the dynamic viscosity and the coefficient of viscosity, is the constant of proportionality. That is:

$$F = \eta A(dv/dr)$$

The unit of viscosity is the pascal second (Pas).

visible radiation *See* light.

visual acuity *See* acuity, visual.

vitreous humour The gelatinous substance behind the lens in the eye. *See* eye.

VLF *See* very low frequency.

volt Symbol: V The SI unit of electrical potential, potential difference, and e.m.f., defined as the potential difference between two points in a circuit between which a constant current of one ampere flows when the power dissipated is one watt. $1 \text{ V} = 1 \text{ J C}^{-1}$.

voltage Symbol: V E.m.f. or electrical potential difference measured in volts.

voltage divider (potential divider) A number of resistors, inductors, or capacitors connected in series with several terminals at intermediate points. The total voltage across the chain can be split into known fractions by making connections at the various points.

voltaic cell *See* cell.

voltameter (coulo(mb)meter) A device for determining electric charge or electric current using electrolysis. The mass m of material released is measured and this can be used to calculate the charge (Q) and the current (I) from the electrochemical equivalent of the element, using $Q = m/z$ or $I = m/zt$. The term now often refers to any case in which two electrodes are immersed in an electrolyte.

voltmeter A meter used to measure electrical voltage. The commonest types are moving-coil instruments with high series resistances to keep the current drawn to a low value. For alternating voltages a rectifier must be used. Moving-iron instruments with high series resistances can be used for both direct and alternating voltages. *See also* potentiometer, electrometer.

volume Symbol: V A measure of the space occupied by an object or system.

volume charge density *See* charge density.

W

watt Symbol: W The SI unit of power, defined as a power of one joule per second. $1 \text{ W} = 1 \text{ J s}^{-1}$.

wattmeter An instrument for measuring power in an electric circuit. The most common type, used in alternating-current circuits, consists of two coils, one fixed and one able to move. The moving coil is connected in series with a high resistance, which is in parallel to the fixed coil and to the load for which the power is being measured. The relative deflection of the coil is proportional to the square of the current through the load; i.e. it is proportional to the power and can therefore be calibrated to give a reading in watts.

wave A method of energy transfer involving some form of vibration. For instance, waves on the surface of a liquid or along a stretched string involve regular to-and-fro motion of particles about a mean position. Sound waves carry energy by alternate compressions and rarefactions of air (or other medium). In electromagnetic waves, electric and magnetic fields vary at right angles to the direction of propagation of the wave.

At any particular instance, a graph of displacement against distance is a regular repeating curve — the *waveform* or wave profile of the wave. In a *travelling* (or *progressive*) *wave* the whole periodic displacement moves through the medium. At any point in the medium the disturbance is changing with time. Under certain conditions a *stationary* (or *standing*) *wave* can be produced in which the disturbance does not change with time.

For the simple case of a plane progressive wave the displacement at a point can be represented by an equation:
$$y = a\sin 2\pi(ft - x/\lambda)$$
where a is the amplitude, f the frequency, x the distance from the origin, and λ the wavelength. Other relationships are:
$$y = a\sin 2\pi(vt - x)/\lambda$$
where v is the velocity, and
$$y = a\sin 2\pi(t/T - x/\lambda)$$
where T is the period. Note that if the $-$ sign is replaced by a $+$ sign in the above equations it implies a similar wave moving in the opposite direction. For a stationary wave resulting from two waves in opposite directions, the displacement is given by:
$$Y = 2a\cos 2\pi x/\lambda$$
See also longitudinal wave, transverse wave, phase.

waveform *See* wave.

wavefront A continuous surface associated with a wave radiation, in which all the vibrations concerned are in phase. A parallel beam has plane wavefronts; the output of a point source has spherical wavefronts.

wave guides *See* microwaves.

wavelength Symbol: λ The distance between the ends of one complete cycle of a wave. Wavelength relates to the wave speed (c) and its frequency (ν) thus:
$$c = \nu\lambda.$$

wave mechanics *See* quantum theory.

wave motion Any form of energy transfer that may be described as a wave rather than as a stream of particles. The term is also sometimes used to mean any harmonic motion.

wave number Symbol: σ The reciprocal of the wavelength of a wave. It is the number of wave cycles in unit distance, and is often used in spectroscopy. The unit is the metre^{-1} (m^{-1}). The circular wave number (Symbol: k) is given by:
$$k = 2\pi\sigma.$$

wave theory of light The theory that energy is transmitted by electromagnetic radiation in the form of waves. The theory is ancient, as is the opposing corpuscular (particle) theory. Tests could not distinguish between them for centuries, until the interference of light was demonstrated in 1801. As this cannot be explained with particle radiation, the problem appeared resolved.

Some decades later, the photoelectric effect was discovered. This cannot be explained by the wave theory. The modern view, based on the quantum theory, is that electromagnetic radiation has a dual nature; it shows wave properties in some situations and particle properties in others.

wavicle A less common name for a photon; used to imply that it has the character of both a wave and a particle.

weak interaction See interaction.

weber Symbol: Wb The SI unit of magnetic flux, equal to the magnetic flux that, linking a circuit of one turn, produces an e.m.f. of one volt when reduced to zero at a uniform rate in one second. 1 Wb = 1 V s.

weight Symbol: W The force by which a mass is attracted to the Earth. It is proportional to the body's mass (m), the constant of proportionality being the gravitational field strength (i.e. the acceleration of free fall). Thus $W = mg$, where g is the acceleration of free fall.

The mass of a body is constant, but its weight varies with position (because it depends on g).

Although mass and weight are often used interchangeably in everyday language, they are different in scientific language and must not be confused.

weightlessness An apparent loss of weight experienced by a person in a spacecraft in orbit. The weight in the Earth's frame of reference is the centripetal force necessary to maintain the circular orbit. In the frame of reference of the spacecraft, the person feels that he has no weight.

Weiss constant See Curie's law.

Weston cadmium cell A standard cell that produces a constant e.m.f. of 1.018 6 volts at 20°C. It consists of an H-shaped glass vessel containing a negative cadmium-mercury amalgam electrode in one leg and a positive mercury electrode in the other. The electrolyte — saturated cadmium sulphate solution — fills the horizontal bar of the vessel to connect the two electrodes. The e.m.f. of the cell varies very little with temperature, being given by the equation:
$$E = 1.018\ 6 - 0.000\ 037\ (T - 293)$$
where T is the absolute temperature.

wet and dry bulb hygrometer See hygrometer.

wet cell A type of cell, such as a car battery, in which the electrolyte is a liquid solution. See also Leclanché cell.

Wheatstone bridge A circuit for measuring electrical resistance. Four resistors, P, Q, R, and S, are connected in a loop (as if along four sides, or 'arms', of a square). A battery is connected across the junctions between P and S and between Q and R. A sensitive galvanometer is connected across the two opposite junctions (P-Q and R-S). If no current flows through the galvanometer, then $P/Q = S/R$. The bridge is then said to be *balanced*.

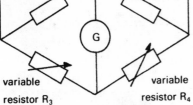

unknown
resistor R_1

standard
resistor R_2

G

variable
resistor R_3

variable
resistor R_4

Wheatstone bridge

If R is an unknown resistance and P, Q, and S are known, then R can be found. P and Q are called the *ratio arms* of the bridge (they determine the ratio of S to R).

There are various ways of using this circuit. In a metre bridge P and Q are a single length of uniform resistance wire with a sliding contact. S is a standard resistor. The position of the contact is moved until the galvanometer gives a zero reading. Then the ratio l_1/l_2 is equal to S/R.

Another method is to use a box containing preset resistances that can be switched to certain values. Typically, the ratio arms, P and R, can each have values of 10, 100, or 1000 ohms, and S can be varied in 1 ohm increments up to 10 000 ohms. Such a device is sometimes called a *Post Office box*.

In Wheatstone-bridge circuits a protective resistance is used with the galvanometer until a near balance is found. It is then shorted out so that a more sensitive balance can be found.

white light The visible radiation produced by a 'white-hot' (incandescent) surface. The temperate must be close to 6000 K for true white light to be produced. White light consists of a continuous range of wavelengths, each of different hue. The sensation of white light can also be produced by suitably mixing hues — two complementary colours, or three primary colours, or more. This effect is 'white' in a physiological rather than a physical sense.

white noise *See* noise.

Wien's displacement law The principle that for black-body radiation:
$$\lambda_{max}T = \text{constant}$$
λ_{max} is the wavelength corresponding to a maximum radiation of energy, and T is the thermodynamic temperature. *See also* black body.

Wimshurst machine A type of electrostatic generator. It consists of two circular discs of insulating material with radial strips of metal foil on their sides. The discs rotate in opposite directions and the charge is produced by friction and collected by metal points.

Wollaston prism A type of polarizing prism. Unlike the Nicol and Rochon prisms, it deviates both the ordinary and the extraordinary rays, by roughly the same angle in opposite directions.

work Symbol: W The work done by a force is the product of the force and the distance moved in the same direction:
work = force × displacement.
Work is in fact a process of energy transfer and, like energy, is measured in joules. If the directions of force (F) and motion are not the same, the component of the force in the direction of the motion is used.
$$W = Fs\cos\theta$$
where s is displacement and θ the angle between the directions of force and motion. Work is the scalar product of force and displacement.

working substance 1. *See* thermometer.
2. The substance undergoing pressure and temperature changes in a heat engine.

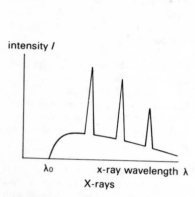

X
intensity *I*

λ₀   x-ray wavelength λ
X-rays

process called bremsstrahlung (braking radiation).

X-ray crystallography The use of X-rays to study crystal structure. The way in which X-rays are diffracted gives information about the spacing between crystal planes.

X-rays Streams of X-radiation.

Y

X-radiation An energetic form of electromagnetic radiation. The wavelength range is 10^{-11} m to 10^{-8} m. X-rays are normally produced by absorbing high-energy electrons in matter. Low-level X-radiation therefore comes from cathode-ray tube screens, including those used in television receivers and computer terminals. The radiation can pass through matter to some extent (hence its use in medicine and industry for investigating internal structures). It can be detected with photographic emulsions and devices like the Geiger–Müller tube.

X-ray photons result from electronic transitions between the inner energy levels of atoms. When high-energy electrons are absorbed by matter, an X-ray line spectrum results. The structure depends on the substance and is thus used in X-ray spectroscopy. The line spectrum is always formed in conjunction with a continuous background spectrum. The minimum (cut-off) wavelength λ_0 corresponds to the maximum X-ray energy, W_{max}. This equals the maximum energy of electrons in the beam producing the X-rays. Wavelengths in the continuous spectrum above λ_0 are caused by more gradual energy loss by the electrons, in the

year The time taken for the Earth to complete one orbit of the Sun. It is measured in various ways. The *solar year* is the time taken for the Sun to make two successive appearances at the point of Aries. It is 365.242 19 mean solar days. The *calendar year* is regulated (using leap years) so that its average length is equal to that of the solar year. The *sidereal year* is measured with respect to the fixed stars. It is 365.256 36 mean solar days. The *lunar year*, which is 12 lunar months, is 365.3671 mean solar days.

yellow spot *See* fovea.

yield point *See* elasticity.

Young modulus *See* elastic modulus.

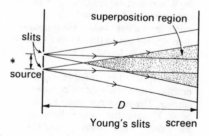

superposition region
slits
* source
D
Young's slits   screen

Young's double slit An experiment to

demonstrate interference of light. Two parallel slits are placed in front of a source of monochromatic light as shown in the illustration. The slits act as two coherent sources of equal intensity and light and dark interference fringes are formed on a screen.

The separation between successive fringes is $\lambda D/2b$, where b is the separation of the slits and D the distance from the slits to the screen. In the original form of the experiment Young used two pinholes, which formed circular fringes.

Z

Zener breakdown *See* Zener diode.

Zener diode A semiconductor diode with high doping levels on each side of the junction. If the junction is reverse-biased, breakdown occurs at a well-defined potential, giving a sharp increase in current. The effect is called *Zener breakdown*; it occurs because electrons are excited directly from the valence band into the conduction band. Zener diodes are used as voltage regulators. *See also* diode, transistor.

zero point energy The energy possessed by the atoms and molecules of a substance at absolute zero (0 K).

APPENDIX

Contents

The Chemical Elements

'Relative atomic mass' in brackets is
nucleon number of most stable isotope.

Element	Symbol	Proton No.	Relative atomic mass
actinium	Ac	89	[227]
aluminium	Al	13	26.9815
americium	Am	95	[243]
antimony	Sb	51	121.75
argon	Ar	18	39.948
arsenic	As	33	74.9216
astatine	At	85	[210]
barium	Ba	56	137.34
berkelium	Bk	97	[247]
beryllium	Be	4	9.0122
bismuth	Bi	83	208.98
boron	B	5	10.81
bromine	Br	35	79.904
cadmium	Cd	48	112.40
caesium	Cs	55	132.905
calcium	Ca	20	40.08
californium	Cf	98	[251]
carbon	C	6	12.011
cerium	Ce	58	140.12
chlorine	Cl	17	35.453
chromium	Cr	24	51.996
cobalt	Co	27	58.9332
copper	Cu	29	63.546
curium	Cm	96	[247]
dysprosium	Dy	66	162.50
einsteinium	Es	99	[254]
erbium	Er	68	167.26
europium	Eu	63	151.96
fermium	Fm	100	[257]
fluorine	F	9	18.9984
francium	Fr	87	[223]
gadolinium	Gd	64	157.25
gallium	Ga	31	69.72
germanium	Ge	32	72.59
gold	Au	79	196.967
hafnium	Hf	72	178.49
helium	He	2	4.0026
holmium	Ho	67	164.930
hydrogen	H	1	1.00797
indium	In	49	114.82
iodine	I	53	126.9044
iridium	Ir	77	192.2
iron	Fe	26	55.847

Element	Symbol	Proton No.	Relative atomic mass
krypton	Kr	36	83.80
lanthanum	La	57	138.91
lawrencium	Lr	103	[257]
lead	Pb	82	207.19
lithium	Li	3	6.939
lutetium	Lu	71	174.97
magnesium	Mg	12	24.305
manganese	Mn	25	54.938
mendelevium	Md	101	[258]
mercury	Hg	80	200.59
molybdenum	Mo	42	95.94
neodymium	Nd	60	144.24
neon	Ne	10	20.179
neptunium	Np	93	[237]
nickel	Ni	28	58.71
niobium	Nb	41	92.906
nitrogen	N	7	14.0067
nobelium	No	102	[255]
osmium	Os	76	190.2
oxygen	O	8	15.9994
palladium	Pd	46	106.4
phosphorus	P	15	30.9738
platinum	Pt	78	195.09
plutonium	Pu	94	[244]
polonium	Po	84	[209]
potassium	K	19	39.102
praseodymium	Pr	59	140.907
promethium	Pm	61	[145]
protactinium	Pa	91	[231]
radium	Ra	88	[226]
radon	Rn	86	[222]
rhenium	Re	75	186.20
rhodium	Rh	45	102.905
rubidium	Rb	37	85.47
ruthenium	Ru	44	101.07
samarium	Sm	62	150.35
scandium	Sc	21	44.956
selenium	Se	34	78.96
silicon	Si	14	28.086
silver	Ag	47	107.868
sodium	Na	11	22.9898
strontium	Sr	38	87.62
sulphur	S	16	32.064

Element	Symbol	Proton No.	Relative atomic mass
tantalum	Ta	73	180.948
technetium	Tc	43	[97]
tellurium	Te	52	127.60
terbium	Tb	65	158.924
thallium	Tl	81	204.37
thorium	Th	90	232.038
thulium	Tm	69	168.934
tin	Sn	50	118.69
titanium	Ti	22	47.90
tungsten	W	74	183.85
uranium	U	92	238.03
vanadium	V	23	50.942
wolfram (tungsten)	W	74	183.85
xenon	Xe	54	131.30
ytterbium	Yb	70	173.04
yttrium	Y	39	88.905
zinc	Zn	30	65.37
zirconium	Zr	40	91.22

Symbols for Physical Quantities

Name of quantity	symbol
absolute temperature	T
absorptance	a
acceleration	a
activity	A
admittance	Y
amount of substance	n
ampere turn	At
angstrom	$\overset{\circ}{A}$
angular acceleration	a
angular displacement	θ
angular frequency	ω
angular magnification	M
angular momentum	L
angular velocity	ω
atomic mass unit	u
Avogadro constant	N
becquerel	Bq
Boltzmann constant	k
Brewster angle	i
bulk modulus	K
calorie	cal
capacitance	C
centripetal acceleration	a
centripetal force	F
characteristic temperature	θ
charge	Q
charge density	ρ
coefficient of friction	μ
concentration	c
conductivity, electrical	σ
conductivity, thermal	λ
cubic expansion coefficient	γ
current	I

Symbols for physical quantities continued

Name of quantity	*symbol*
current density	j
debye	D
decay constant	λ
decibel	dB
declination	ε
degeneracy of an energy level	g
density	ρ
diameter	d
diffusion coefficient	D
displacement	s
efficiency	η
electric charge	Q
electric displacement	D
electric field strength	E
electric flux	ψ
electric potential	V
electrochemical equivalent	Z
electromotive force	E
electronvolt	eV
emissivity	ε
energy	W
enthalpy	H
entropy	S
exitance	M
Faraday constant	F
foot	ft
force	F
frequency	f, ν
gas constant	R

Symbols for physical quantities continued

Name of quantity	*symbol*
Gibbs function	G
gram	g
gravitational constant	G
half-life	$T_{\frac{1}{2}}$
heat	Q
height	h
Helmholtz function	F
horizontal intensity	B
horsepower	hp
illumination	E
impedance	Z
inclination	δ
internal energy	U
ionization potential	I
irradiance	E
Joule's equivalent	J
kilowatt-hour	kwh
kinematic viscosity	ν
kinetic energy	T
light-year	ly
litre	l
luminance	L_v
luminous flux	ϕ_v
luminous intensity	I_v
magnetic field strength	H

Symbols for physical quantities continued

Name of quantity	symbol
magnetic flux	ϕ
magnetic flux density	B
magnetic moment	m
magnetomotive force	F
magnification	m
mass	m
maxwell	Mx
mean free path	λ
mean life	T
molar thermal capacity	C_m
moment of inertia	I
momentum, linear	p
multiplication factor	k

Néel temperature	θ
neper	N_p
neutron number	N
nit	nt
nucleon number	A

oersted	Oe
optical path	d

parity	P
parsec	pc
period	T
permeability	μ
permittivity	ε
phon	p
Planck constant	h
poise	P
Poisson ratio	ν
potential difference	V
potential energy	V

Symbols for physical quantities continued

Name of quantity	symbol
poundal	pdl
power	P
Poynting vector	S
pressure	p
proton number	Z
radiance	L_e
radiant flux	Ω_e
radiant intensity	I_e
radioactivity	A
radius	r
radius of curvature	r
radius of gyration	k
reactance	X
reflectance	ρ
refractive constant	n
relative atomic mass	A_r
relative density	d
relative molecular mass	M_r
relative permittivity	E_r
reluctance	R
resistance	R
resistivity	ρ
restitution, coefficient of	e
rest mass	m_0
Reynolds number	Re
roentgen	R
shear modulus	G
solid angle	Ω
specific rotatory power	a_m
speed	c
stokes	St
surface tension	r, o
susceptibility	X

Symbols for physical quantities continued

Name of quantity	symbol
temperature	T, t
thermal capacity	C
thermodynamic temperature	T
time	t
tonne	t
torque	T
velocity	v
viscosity	η
voltage	V
volume	V
wavelength	λ
wave number	σ
weight	W
work	W
Young modulus	E

Conversion Factors

SI, c.g.s, and f.p.s. (foot-pound-second) units

Length

	m	cm	in	ft	yd
1 metre	1	100	39.3701	3.28084	1.09361
1 centimetre	0.01	1	0.393701	0.0328084	0.0109361
1 inch	0.0254	2.54	1	0.0833333	0.0277778
1 foot	0.3048	30.48	12	1	0.0333333
1 yard	0.9144	91.44	36	3	1

	km	mi	n.mi
1 kilometre	1	0.621371	0.539957
1 mile	1.60934	1	0.868976
1 nautical mile	1.85200	1.15078	1

1 light year $= 9.46070 \times 10^{15}$ metres $= 5.87848 \times 10^{12}$ miles
1 Astronomical Unit $= 1.495 \times 10^{11}$ metres.
1 parsec $= 3.0857 \times 3.2616$ light years.

Conversion factors continued

Mass

	kg	g	lb	long ton
1 kilogram	1	1000	2.20462	9.84207×10^{-4}
1 gram	10^{-3}	1	2.20462×10^{-3}	9.84207×10^{-7}
1 pound	0.453592	453.592	1	4.46429×10^{-4}
1 long ton	1016.047	1.016047×10^{6}	2240	1

Velocity

	m/sec	km/hr	mi/hr	ft/sec
1 metre per second	1	3.6	2.23694	3.28084
1 kilometre per hour	0.277778	1	0.621371	0.911346
1 mile per hour	0.44704	1.609344	1	1.46667
1 foot per second	0.3048	1.09728	0.681817	1

Force

	N	kg	dyne	poundal	lb
1 newton	1	0.101972	10^{5}	7.23300	0.224809
1 kilogram force	9.80665	1	9.80665×10^{5}	70.9316	2.20462
1 dyne	10^{-5}	1.01972×10^{-6}	1	7.23300×10^{-5}	2.24809×10^{-6}
1 poundal	0.138255	1.40981×10^{-2}	1.38255×10^{4}	1	0.031081
1 poundal force	4.44822	0.453592	4.44823×10^{5}	32.174	1

Conversion factors continued

Pressure

	N/m^2 (Pa)	kg/cm^2	lb/in^2	atmos
1 newton per square metre (pascal)	1	1.01972×10^{-5}	1.45038×10^{-4}	9.86923×10^{-6}
1 kilogram per square centimetre	980.665×10^2	1	14.2234	0.967841
1 pound per square inch	6.89476×10^3	0.0703068	1	0.068046
1 atmosphere	1.01325×10^5	1.03323	14.6959	1

1 pascal=1 newton per square metre=10 dynes per square centimetre.
1 bar=10^5 newtons per square metre=0.986 923 atmosphere.
1 torr=133.322 newtons per square metre=1/760 atmosphere.
1 atmosphere=760mm Hg=29.92in Hg=33.90ft water (all at 0°C.).

Work and Energy

	J	cal_{IT}	kWhr	btu_{IT}
1 joule	1	0.238846	2.77778×10^{-7}	9.47813×10^{-4}
1 calorie$_{(IT)}$	4.1868	1	1.16300×10^{-6}	3.96831×10^{-3}
1 kilowatt hour	3.6×10^6	8.59845×10^5	1	3412.14
1 British Thermal Unit$_{(IT)}$	1055.06	251.997	2.93071×10^{-4}	1

1 joule=1 newton metre=1 watt second=10^7 ergs=0.737 561ft lb.
1 electron volt=$1.602 \ 10 \times 10^{-19}$joule.

Conversion factors continued

Density

	kg m^{-3}	g cm^{-3}	lb ft^{-3}	lb in^{-3}
1 kilogramme per cubic metre	1	10^{-3}	0.062 428	3.612 73×10^{-5}
1 gramme per cubic centimetre	1000	1	62.428	3.612 73×10^{-2}
1 pound per cubic foot	16.0185	0.0160185	1	5.787 04×10^{-4}
1 pound per cubic inch	2.767 99×10^4	27.6799	1728	1

1 lb/gal (UK)=0.099 776 3 kg dm^{-3}

Volume

	m^3	cm^3	in^3	ft^3	gal
1 cubic metre	1	10^6	6.102 36×10^4	35.3146	219.969
1 cubic centimetre	10^{-6}	1	0.061 023 6	3.531 46×10^{-5}	2.199 69×10^{-4}
1 cubic inch	1.638 71×10^{-5}	16.3871	1	5.787 04×10^{-4}	3.604 64×10^{-3}
1 cubic foot	0.028 316 8	28 316.8	1728	1	6.228 82
1 gallon (UK)	4.546 09×10^{-3}	4546.09	277.42	0.160544	1

1 gallon (US)=0.832 68 gallon (UK)

The *litre* is now recognized as a special name for a cubic decimetre, but is not used to express high precision measurements.

Conversion factors continued

Area

	m²	cm²	in²	ft²
1 square metre	1	10^4	1550	10.7639
1 square centimetre	10^{-4}	1	0.155	1.07639×10^{-3}
1 square inch	6.4516×10^{-4}	6.4516	1	$6.944 44 \times 10^{-3}$
1 square foot	9.2903×10^{-2}	929.03	144	1

	m²	km²	yd²	mi²	acre
1 square metre	1	10^{-6}	1.195 99	$3.860 19 \times 10^{-7}$	$2.471 05 \times 10^{-4}$
1 square kilometre	10^6	1	1.195×10^6	0.386019	247.105
1 square yard	0.836 127	$8.361 27 \times 10^{-7}$	1	$3.228 31 \times 10^{-7}$	$2.066 12 \times 10^{-4}$
1 square mile	$2.589 99 \times 10^6$	2.589 99	3.0976×10^6	1	640
1 acre	$4.046 86 \times 10^3$	$4.046 86 \times 10^{-3}$	4840	1.5625×10^{-3}	1

1 are=100 square metres.
1 hectare=10 000 square metres=2.471 05 acres.

The Greek Alphabet

Letters		Name
A	α	alpha
ß	β	beta
Γ	γ	gamma
Δ	δ	delta
E	ε	epsilon
Z	ζ	zeta
H	η	eta
Θ	θ	theta
I	ι	iota
K	κ	kappa
Λ	λ	lambda
M	μ	mu
N	ν	nu
Ξ	ξ	xi
O	o	omicron
Π	π	pi
P	ρ	rho
Σ	σ	sigma
T	τ	tau
Υ	ν	upsilon
Φ	φ	phi
X	χ	chi
Ψ	ψ	psi
Ω	ω	omega